AF725515

Understanding Agroecology

Understanding Agroecology

Layla Warner

Callisto Reference

www.callistoreference.com

Callisto Reference,
118-35 Queens Blvd., Suite 400,
Forest Hills, NY 11375, USA

Visit us on the World Wide Web at:
www.callistoreference.com

© Callisto Reference, 2022

This book contains information obtained from authentic and highly regarded sources. All chapters are published with permission under the Creative Commons Attribution Share Alike License or equivalent. A wide variety of references are listed. Permissions and sources are indicated; for detailed attributions, please refer to the permissions page. Reasonable efforts have been made to publish reliable data and information, but the authors, editors and publisher cannot assume any responsibility for the validity of all materials or the consequences of their use.

ISBN: 978-1-64116-520-4 (Hardback)

Trademark Notice: Registered trademark of products or corporate names are used only for explanation and identification without intent to infringe.

Cataloging-in-Publication Data

Understanding agroecology / Layla Warner.
p. cm.
Includes bibliographical references and index.
ISBN 978-1-64116-520-4
1. Agricultural ecology. 2. Agriculture--Environmental aspects.
3. Agriculture. 4. Ecology. I. Warner, Layla.
S589.7 .U53 2022
577.55--dc23

Table of Contents

Preface VII

Chapter 1 Introduction 1
- Agroecology 1
- Elements of Agroecology 2
- Agroecology, Food Sovereignty and the New Green Revolution 4
- Agroecology and Sustainable Food Production System 8
- Sustainable Land Management 9

Chapter 2 Agroecosystem 16
- Agricultural Biodiversity 17
- Ecological Restoration of Farmland 23
- Agroecosystem Analysis 38
- Microbial Impact on Agroecosystem 40
- Soil Organic Matter 48
- Agroecosystem Management 51
- Challenges to Agroecosystem Management 57

Chapter 3 Stresses in Agroecology 65
- Biotic Stress 66
- Abiotic Stress 69
- Interaction of Plant Biotic and Abiotic Stresses 87
- Agriculture Biotechnology for Biotic and Abiotic Stress 102

Chapter 4 Pest and Weed Management 110
- Ecological Pest Management 110
- Bioeffector 113
- Biological Pest Control 115
- Biopesticide 128
- Biological Weed Control 138

Chapter 5 Allied Fields of Agroecology 147

- Sustainable Agriculture 147
- Agroforestry 151
- Organic Farming 153
- Carbon Farming 186
- Sustainable Gardening 188
- Permaculture 192
- Polyculture 201
- Silviculture 206

Permissions

Index

Preface

This book is a culmination of my many years of practice in this field. I attribute the success of this book to my support group. I would like to thank my parents who have showered me with unconditional love and support and my peers and professors for their constant guidance.

The study of ecological processes which are used in agricultural production systems is referred to as agroecology. It applies the principles of ecology to the management of agroecosystems. It also focuses on the basic properties of agroecosystems such as sustainability, stability, productivity and equitability. Some of the major varieties of agroecology are ecosystem agroecology and agro-population ecology. Agroecology is an interdisciplinary subject which uses various sciences to examine the elements of agroecosystems such as soil properties and the interactions between plants and insects. It also utilizes social sciences to study how rural communities get affected by farming practices. This book is a compilation of chapters that discuss the most vital concepts in the discipline of agroecology. Most of the topics introduced in this book cover new techniques and applications of this field. It aims to serve as a resource guide for students and experts alike and contribute to the growth of agroecology.

The details of chapters are provided below for a progressive learning:

Chapter - Introduction

The study of ecological processes applied to agricultural production systems is called agroecology. Some of its important elements are synergies, efficiency, recycling, resilience, and responsible governance. This chapter closely examines these elements of agroecology as well as sustainable land management to provide an extensive understanding of the subject.

Chapter - Agroecosystem

Agroecosystem is an ecosystem on agricultural land that includes the diverse living and non-living components as well as their interactions. Some of the microbial impacts on agroecosystem include nutrient cycling and nitrogen fixation. The topics elaborated in this chapter will help in gaining a better perspective about agroecosystem as well as their management.

Chapter - Stresses in Agroecology

Stresses in agroecology are categorized as biotic and abiotic stresses. Some of the varied aspects of agroecology stress include interaction of plant biotic and abiotic stress as well as use of agriculture biotechnology for management of these stresses. This chapter discusses in detail these environmental stresses related to agroecology.

Chapter - Pest and Weed Management

The approach to pest and weed regulation which involves increasing the strengths of natural systems to reinforce the natural processes is referred to as ecological pest and weed management. It involves the use of bioeffectors, bioherbicides and biopesticides. These aspects of ecological pest management have been thoroughly discussed in this chapter.

Chapter – Allied Fields of Agroecology

There are various fields which are closely associated with agroecology such as agroforestry, sustainable agriculture, carbon farming, sustainable gardening, permaculture, polyculture, silviculture and organic farming. The topics elaborated in this chapter will help in gaining a better perspective about these allied fields of agroecology.

Layla Warner

1

Introduction

The study of ecological processes applied to agricultural production systems is called agroecology. Some of its important elements are synergies, efficiency, recycling, resilience, and responsible governance. This chapter closely examines these elements of agroecology as well as sustainable land management to provide an extensive understanding of the subject.

AGROECOLOGY

Agroecology is 'the application of ecological concepts and principles to the design and management of sustainable agro-ecosystems' – has three facets. It is:

1. A scientific discipline involving the holistic study of agro-ecosystems, including human and environmental elements.
2. A set of principles and practices to enhance the resilience and ecological, socio-economic and cultural sustainability of farming systems.
3. A movement seeking a new way of considering agriculture and its relationships with society.

A growing body of evidence reveals agroecology's multiple advantages over conventional high-external input farming:

A multi-functional approach to farming, capable of meeting environmental, economic and social needs greater environmental sustainability and resilience, especially in marginal areas subject to environmental degradation and extreme climatic events, and higher agrobiodiversity the ability to support farmers' food sovereignty, reducing their dependence on costly and sometimes difficult-to-access chemical inputs higher overall productivity (at farm rather than crop level) achieved through a diverse range of agricultural products and environmental services, which reduce risks of crop failure in the long term.

Barriers to Agroecology

Agroecology is knowledge- and management-intensive. A common deterrent to its adoption is the large amount of 'soft' inputs (such as knowledge and skills) and labour that some practices require in the initial seasons. Poorer and more marginal farmers, in particular, may decide not to adopt these practices if they do not have enough time and resources to invest in learning and experimentation.

Policies and market signals are stacked against agroecology. The prices of conventional agricultural products are distorted by heavy subsidies – both direct (such as farm and input subsides) and indirect (the health and environmental consequences of unsustainable practices that are paid for by taxpayers).

Insecure land tenure and lack of access to natural resources can also inhibit the spread of agroecology by discouraging farmers from adopting practices that require long-term investment in land and other assets.

Agroecology is viewed as 'alternative' and sometimes in direct opposition to conventional farming. Yet the differences are often blurred. Emphasising the two approaches as mutually-exclusive may undermine the constructive search for more common ground.

Doing the Needful

Consolidate the evidence base to support agroecology through multi-dimensional analyses that provide comparable measures not just of crop yields, but also non-commodity outputs, such as feed available for livestock, mulching crops, provision of ecosystem services and contributions to the incomes and resilience of farming households.

Promote a fundamental cultural and philosophical shift in the evaluation of what is regarded as 'productive' and 'efficient', not just by farmers, but by society as a whole. The emphasis should be on optimising rather than maximising production (and profits). This is probably the most challenging step required to mainstream agroecological principles into agricultural policies and practices.

Assess the constraints to the adoption of agroecological practices and the policy options to overcome them. These may include innovative tools and approaches such as participatory innovation systems, economic incentives for early adopters, payments for environmental services and rewards for landscape conservation. Relevant policy changes may also concern land tenure, natural resources management as well as support to farmer organisations, local business development and markets for agroecological products.

Reduce the polarisation between agroecology and external input-intensive agriculture. Encouraging the adoption of aspects of agroecology by the latter could help reduce the negative environmental and social impacts of this production model, and the first step towards greater sustainability. On the other hand, selected 'modern' technologies such as precision farming and new breeding techniques can help agroecological farmers to respond to new challenges posed by the changing climate, water stress, land degradation and fast cultural and socio-economic transformations.

ELEMENTS OF AGROECOLOGY

Elements of agroecology are meant to guide us toward the transition to sustainable food and agricultural systems. These interlinked and interdependent elements are as follows:

- Diversity: By using varied agricultural production systems such as agroforestry (incorporating trees into farming systems) or polycultures (a wide variety of crops in the same space), agroecology contribute to a range of production, socio-economic, nutrition and environmental benefits.

- Co-creation and sharing of knowledge: Agroecology depends on context-specific knowledge. Knowledge plays a central role in the process of developing and implementing agroecological innovations to address challenges across food systems. Through the co-creation process, agroecology blends global scientific data with the traditional, indigenous, practical and local knowledge of producers.

Agroecology blends global scientific data with the traditional, indigenous, practical and local knowledge of food producers.

- Synergies: When designing an agricultural system, all of its aspects, such as crops, animals, trees, soils and even community involvement, should be taken into consideration. Creating synergies between pieces of a system helps them to function better, leading to improved soil fertility, natural pest regulation and increased crop productivity.

- Efficiency: Efficiency starts with not wasting resources. By more efficiently using inputs to agriculture (such as seeds, soil, energy, nutrients), agroecology uses fewer external resources, reducing costs and negative environmental impacts. This has the knock-on effect of conserving precious resources like water, protecting biodiversity and even reducing the costs of production.

- Recycling: Nature reuses what it produces. By imitating natural ecosystems, agroecological practices support biological processes that drive the recycling of nutrients, biomass and water within production systems. This process of recycling can be enhanced by certain actions, such as introducing livestock and using manure as fertilizer as one example; however, recycling at all levels is key to self-sustaining, self-correcting systems.

- Resilience: Enhancing ecological and socio-economic resilience, agroecological systems have a greater capacity to recover from disasters such as drought, floods or hurricanes, and to resist pest and disease attack. Through diversification, producers reduce their vulnerability in the case that a single crop or commodity fails. Reducing dependence on external inputs increases producers' autonomy and reduces their vulnerability to economic risk.

- Human and social values: Agroecology places a strong emphasis on human and social values, such as dignity, equity, inclusion and justice, all contributing to sustainable livelihoods. It puts the aspirations and needs of those who produce, distribute and consume food at the heart of food systems. Agroecology seeks to address inequalities by creating opportunities for women and youth.

- Culture and food traditions: Agriculture is part of humankind's heritage, with food traditions playing a central role in society. However, in many places, there is a disconnection between food habits and culture. This has contributed to a situation where hunger and obesity exist side by side. Agroecology plays an important role in reconciling tradition and modern food habits, bringing them together in a harmonious way that promotes local, seasonal food and healthy, diversified and culturally-appropriate diets that deliver good nutrition while reducing the food industry's carbon footprint and protecting ecosystems.

Agroecology can help us transition to sustainable food and agricultural systems.

- Responsible governance: Transparent, accountable and inclusive governance mechanisms on different scales are necessary to create an enabling environment that supports producers to transform their systems. Equitable access to land and natural resources is not only key to social justice, but also essential to providing incentives for long-term investments in sustainability.

- Circular and solidarity economy: Local solutions are at the core of agroecology. This includes supporting local markets and economies that offer fair and sustainable livelihoods to its community members. Agroecology seeks to shorten food circuits by decreasing the number of intermediaries, thereby increasing the incomes of food producers while keeping prices fair for consumers.

AGROECOLOGY, FOOD SOVEREIGNTY AND THE NEW GREEN REVOLUTION

The global food crisis of returned in with devastating impacts on the world's poor—most of whom are peasant farmers. Hunger resulted not from a lack of global food stocks but from food price inflation. Volatility and high food prices have led institutions in the corporate food regime to call for a 70% increase in food production by 2050.

A food regime is a "rule governed structure of production and consumption of food on a world scale". The present corporate food regime is made up of the global food system's government ministries, global institutions, agri-food monopolies, land grant universities, think tanks, and big philanthropy that generate the technologies, the discourse, and enforce the regime's "rules" (e.g., free trade agreements, the U.S. Farm Bill and the European Common Agricultural Policy [CAP]).

With the food crisis, international institutions produced a steady stream of assessments calling for investment in biotechnology and a new Green Revolution. With the notable exception of the International Assessment of Agricultural Knowledge, Science and Technology for Development (IAASTD)—these reports rest on several problematic suppositions: that grain-fed meat consumption will expand in emerging economies; that arable land will be diverted to agro-fuels; that financial speculation and price volatility in food commodities will continue unchecked; that production increases depend on transgenic, proprietary technologies and external inputs; and that liberalized, global trade is essential to food security.

These suppositions buttress the political-economic assertion behind the 70% by 2050 call: It is not proprietary, Green Revolution agriculture and liberalized global markets that have caused the food crises per se, but their inefficient or inadequate application. Therefore, the solution is to do more of the same, over a greater area, more efficiently. As Amin indicates, this neoliberal strategy is:

> The "absolute and superior rationale" of economic management based on the private and exclusive ownership of the means of production. According to this principle, land and labor become merchandise like any other commodity, and are transferable at the market price in order to guarantee their best use for their owners and for society as a whole. This is nothing but a mere tautology, yet it is the one upon which all critical economic discourse is based.

Faced with stagnant global economic growth, this paradigm views the peasantry as a site for "accumulation by dispossession", and as a sector for potential market expansion. Because their numbers are growing at 8% a year, market access to the 2.5 billion farmers at the "base of the pyramid" has become attractive for global capital.

As in the 1960–1980s, capital's key to the peasantry's land, factor, and commodity markets is, once again, the Green Revolution. Similar to the role once played by the Ford and Rockefeller foundations, the Bill and Melinda Gates Foundation is the Green Revolution's new philanthropic flagship, tasked with resurrecting the Consultative Group on International Agricultural Research (CGIAR) and obtaining broad social, financial, and government agreement. The new "Doubly Green Revolution", retains the same proprietary genetic foundations as the original Green Revolution, but has added transgenic technologies, global markets, environmental concerns, and a leading role for the private sector. The U.S. Agency for International Development's Feed the Future, the Gates Foundation's Alliance for a Green Revolution in Africa (AGRA), and industry's New Vision for African Agriculture initiatives, for example, feature value chains, public-private partnerships, micro-finance, village "agro-dealers" and smallholder contract farming.

Widespread social, environmental, and agricultural critiques of the Green Revolution notwithstanding, food regime institutions have steadily converged around the new Green Revolution agenda.

Peasant Agriculture and Agroecology: A Means and a Barrier for the Green Revolution

The planet's smallholders and the practice of agroecology both constitute a means and a barrier to the expansion of capitalist agriculture. Smallholders subsidize capitalist agriculture with cheap labor and supply a vast, low-end factor market. This functional dualism between peasant and capital-intensive agriculture accelerates industrial expansion, resulting in the differentiation and displacement of the peasantry and the subsumption of peasant agriculture to capitalist agriculture. At the same time, family labor, small farm size, diversified farming and knowledge systems, and smallholder's pluriactive livelihood strategies preserve peasant farming systems, presenting barriers and competition for capitalist agriculture.

Traditional agriculture was the cultural and ecological basis for the development of agroecology as a science. Because it is rooted in smallholder systems and relies on agroecosystem management rather than external inputs, agroecology is also a barrier to Green Revolution technologies. Agroecology is knowledge intensive (rather than capital intensive), tends toward small, highly diversified farms, and emphasizes the ability of local communities to generate and scale-up innovations through farmer-to-farmer research and extension approaches.

The first Green Revolution drew in millions of smallholders, many of whom were forced out of farming by larger, better capitalized farmers, or went bankrupt after their soils became sterile and subsidized credit disappeared. Over 70% of the world's agrobiodiversity—largely held in situ in smallholder agroecosystems—was lost from farming. When smallholder farms began crashing under Green Revolution methods in the 1970s, many farmers turned to agroecology in an effort to restore soil organic matter, conserve water, restore agrobiodiversity, and manage pests. Since the early 1980s, hundreds of nongovernmental organizations (NGOs) in Africa, Latin America, and Asia have promoted thousands of agroecology projects that incorporate elements of traditional knowledge and modern agroecological science. With the growing food, fuel, and climate crises, the importance of the ecological and social services provided by agroecological peasant agriculture are becoming widely recognized.

In Latin America, the expansion of agroecology has produced cognitive, technological, and sociopolitical innovations, intimately linked to new political scenarios such as the emergence of progressive governments in Ecuador, Bolivia, and Brazil, and peasants/indigenous resistance movements. Thus, agroecology's "epistemological, technical and social revolution" is mutually constitutive with social movements and political processes.

While the Green Revolution has been "greening" itself since its highly publicized renewal, its champions have criticized agroecology's alleged low productivity and for not "scaling up." These criticisms ignore the evidence demonstrating the high productivity and resilience of agroecologically managed peasant agriculture, and forget that scaling up the first Green Revolution required the massive structural mobilization of state and private-sector resources.

While agroecology has spread widely through the efforts of NGOs, farmers' movements and university projects, it remains marginal to official agricultural development plans and is dwarfed by the resources provided to the Green Revolution. In contrast, the remarkable scaling up of agroecology in Cuba stems, in large part, from the government's strong structural support. Asking "Why can't agroecology scale up?" begs the question, "What is holding agroecology back?"

Green Revolution and Agroecology: Marriage or Functional Dualism

Given its popularity and its potential, some governments, universities, and even big philanthropy are selectively incorporating technical aspects of agroecology that do not challenge the politics of the

Green Revolution. Some organic farmers and ecologists suggest that a marriage between agroecology, organic farming, and biotechnology can close yield gaps while reducing industrial agriculture's environmental footprint through sustainable intensification, for example, increasing efficiency of inputs and deploying climate-smart genetic varieties. The Gates Foundation is adding-on integrated soil fertility management to its. Advocates for these approaches suggest that because of the severity of the food crisis, we need all solutions that is, productive genetically modified organisms (GMOs) and unproductive (but greener) agroecological practices. Invariably, agroecology receives a fraction of the funding provided to Green Revolution technologies. Agroecology is further subordinated to conventional agriculture by revisionist academic projects that erase its history, stripping it of its political content. By co-opting agroecology, relegating it to the margins of science and niche markets of the corporate food regime, these strategies advance a form of "functional dualism".

Agroecology and Food Sovereignty Movements

A new Green Revolution could conceivably concentrate food production on some 50,000 industrial farms worldwide. Given the best land, subsidized inputs, and favorable market access, these farms could produce the world's food (although not very sustainably). But how would 2.5 billion displaced smallholders buy this food? The alternative—smallholderdriven agroecological agriculture—was recognized by the scientists of the International Agricultural Assessment of Knowledge Science and Technology for Development (IAASTD) as the best strategy for rebuilding agriculture and ending rural poverty and hunger,

The wealth of agricultural knowledge, science, and technology (AKST) the world has built up should be targeted toward agroecology strategies that combine productivity with protecting natural resources like soils, water, forests, and biodiversity. In particular, the research and development efforts must now target and include in a participatory manner small-scale and family farmers, since they make up the major part of the poor and hungry, while they also represent the major part of the stewards for the environment. Agricultural practices like organic, biodynamic, conservation, and agroecological are options that address the main constraints to food and nutrition security as well as food sovereignty issues.

To be an effective strategy, major changes must be made in policies, institutions, and research priorities to create an enabling environment for peasant-based, agroecological development. This transformation will likely require a combination of extensive on-the-ground agroecological practice and strong political will to overcome opposition and co-optation from the Green Revolution.

Smallholders working with movements like Campesino a Campesino (Farmer to Farmer) of Latin America, and NGO networks for farmer-led sustainable agriculture like Participatory Land Use Management (PELUM) of Africa, have restored degraded farmland using highly effective agroecological practices on hundreds of thousands of acres of land.

At the same time, peasant organizations fighting for agrarian reform have confronted commodity dumping, market-based land reform and more recently, extensive land grabs. The international peasant federation La Via Campesina has called for food sovereignty, "The right of people's to healthy and culturally appropriate food, produced through ecologically sound and sustainable methods, and their right to define their own food and agricultural systems". The cross-border globalization of these movements responds in part to the intensification of capital's enclosures and is partly a strategic decision to engage in global advocacy.

The need for structural support for smallholders in locally based agroecology networks, and the globalized agrarian demands of the food sovereignty movement are complementary areas of strategic synergy. The food crisis is drawing them toward convergence.

When NGO federation PELUM brought over 300 farmer-extensionists to Johannesburg to speak on agroecology at the World Summit on Sustainable Development, farmers formed the Eastern and Southern Africa Farmers Forum to address agrarian issues. Following the Rome food crisis meeting in 2008, La Via Campesina met in Mozambique where they signed a declaration for a smallholder's agroecological solution to the food crisis. Developments like this (and many others) suggest that the international call for food sovereignty is beginning to take root in smallholder agroecology networks. Similarly, La Via Campesina is steadily spreading agroecological approaches throughout its own farmer organizations.

As local networks for agroecological practice merge with the transnational agrarian movements for food sovereignty they generate massive social pressure—pressure that is needed to tip the scales of political will in favor of food sovereignty and agroecology. This pressure can take the form of constitutional reform, for example, Ecuador's food sovereignty law, grassroots campaigns and civil society declarations linking agroecological practice to political practice or the adoption of agroecology as a development strategy for example Brazilian Landless Worker's Movement's schools and training programs.

However, this convergence faces historical divisions between agrarianbased farmer organizations and the NGO-based agroecology networks. The latter are more easily co-opted into technical and apolitical approaches to agricultural development. This has led long-time agroecology practitioners to call for a shift in NGO behavior and priorities, from technology-led agendas, to strategies that support farmer-led political organizations.

The call for strategic alliances also comes from peasant leaders. Alberto Gómez of Mexico's National Union of Peasant Organizations (UNORCA) affirms, "We have to form alliances with technicians or with NGOs that complement our activities our struggle is not only in the political arena, in movement building, it's also about building local alternatives. It is about creating a different context for agriculture and peasant life. In this sense there are complementarities".

AGROECOLOGY AND SUSTAINABLE FOOD PRODUCTION SYSTEM

Agroecology uses disciplines from modern agricultural science but its approach is also influenced by the indigenous knowledge systems about soils, plants, etc. that have nurtured traditional farming systems for millennia. Agroecology does not promote technical recipes but rather principles, it is not an agriculture of inputs but rather of processes.

In order for the technologies derived from the application of principles to be relevant to the needs and circumstances of small farmers, the technological generation process must result from a participatory research process in which farmers along with researchers provide input into the research questions and the design, running and evaluation of field experiments. Besides science and practice, agroecology also refers to a wide variety of social movements aimed at environmental protection, the development of sustainable farming systems and food sovereignty.

The concept of "movement" is used in order to stress the vision of social and economic positive impact that Agroecology potentially has for sustainable rural development The Agroecology sector emerged as a different model to address the problems of world agriculture caused by industrial food production model and it is based on the principles of Sustainability, Integrity, Equality, Performance and Stability.

The primary concepts of agroecology and corresponding management practices resonate with arguments for food security, food sovereignty and sustainable rural development. The agroecological concepts and principles embrace also a wide range of practices and have broad scope for implementation. This means that they have considerable resonance with other concepts, principles and practices in the field of sustainable agriculture that also offer alternative structures to the mainstream paradigm of industrial agriculture. Such key approaches are permaculture, biodynamic agriculture, organic farming, conservation agriculture, urban farming, and multifunctionality in agriculture.

The context of agroecology expanded over the years from the field, through the farm, to the landscape; and in a wider understanding to the food system level. Agroecology promotes ecologically and culturally sound food systems and food sovereignty, protecting people's ability and right to define their own models of food production, distribution and consumption. Such an approach in rural development can contribute to empowering disadvantaged communities through diversifying activities of farmers and including new groups in different levels of the food system-production, processing and trade. This helps to strengthen employment; local food security and food and prevent population decrease.

SUSTAINABLE LAND MANAGEMENT

Land provides an environment for agricultural production, but it also is an essential condition for improved environmental management, including source/sink functions for greenhouse gasses, recycling of nutrients, amelioration and filtering of pollutants, and transmission and purification of water as part of the hydrologic cycle. The objective of sustainable land management (SLM) is to harmonise the complimentary goals of providing environmental, economic, and social opportunities for the benefit of present and future generations, while maintaining and enhancing the quality of the land (soil, water and air) resource. Sustainable land management is the use of land to meet changing human needs (agriculture, forestry, conservation), while ensuring long-term socioeconomic and ecological functions of the land.

Sustainable land management is a necessary building block for sustainable agricultural development, and it is a key element in AGENDA 21's goal of sustainable development. Sustainable agricultural development, conservation of natural resources, and promoting sustainable land management are key objectives of the new World Bank rural investment program, From Vision to Action, and increasingly these objectives are being included in all agricultural development and natural resources management projects.

Sustainable land management combines technologies, policies, and activities aimed at integrating socioeconomic principles with environmental concerns, so as to simultaneously:

- Maintain and enhance production (productivity).
- Reduce the level of production risk, and enhance soil capacity to buffer against degradation processes (stability/resilience).
- Protect the potential of natural resources and prevent degradation of soil and water quality (protection).
- Be economically viable (viability).
- Be socially acceptable, and assure access to the benefits from improved land management (acceptability/equity).

The definition and these criteria, called pillars of SLM, are the basic principles and the foundation on which sustainable land management is being developed. Any evaluation of the sustainability has to be based on these objectives: productivity, stability/resilience, protection, viability, and acceptability/equity. The definition and pillars have been field tested in several countries, and they were judged to provide useful guidance to assess sustainability.

The lack of a comprehensive, quantifiable definition for sustainable land management is sometimes considered to be a serious deficiency. Yet, as argued by, a research model for sustainability has to be more flexible and therefore less easy to quantify than a research model for chemistry, physics, or classical agronomy. Such a research model must be designed around an evaluation process (rather than thematic context), because it is intended to test the likelihood of certain events taking place and the aggregate impacts of these events, rather than specifics of various null hypothesis or the impacts of certain inputs or land management interventions. Essentially the research model must include a goal statement, a conceptual framework, a set of procedures, and criteria (indicators) for diagnosis. One of the main objectives of such a research model is to evaluate the impacts of events which are uncertain, but the process of evaluation is guided by scientifically defined protocols.

Principles and Criteria for Sustainable Land Management

Experiences gained from field projects in developing and developed countries has identified a series of principles for sustainable land management, and these can be used as general guidelines for development projects. The most useful of these are summarized below:

Global concerns for sustainability:

- Sustainability can be achieved only through the collective efforts of those immediately responsible for managing resources. This requires a policy environment that empowers farmers and other, local decision makers, to reap benefits for good land use decisions, but also to be held responsible for inappropriate land uses.

- Integration of economic and environmental interests in a comprehensive manner is necessary to achieve the objectives of sustainable land management. This requires that environmental concerns be given equal importance to economic performance in evaluating the impacts of development projects, and that reliable indicators of environmental performance be developed.
- There is urgent need to resolve the global challenge to produce more food to feed rapidly rising global populations, while at the same time preserving the biological production potential, resilience, and environmental maintenance systems of the land. Sustainable land management, if properly designed and implemented, will ensure that agriculture becomes a part of the environmental solution, rather than remaining an environmental problem.

Sustainable Agriculture:

- More ecologically balanced land management can achieve both economic and environmental benefits, and this must be the foundation (linch pin) for further rural interventions (investments). Without good land management, other investments in the rural sector are likely to be disappointing. At the same time, arguing for the continued maintenance of agriculture without reference to environmental sustainability is increasingly difficult. Indicators of land quality are needed to guide us along the way.
- Agricultural intensification is often necessary to achieve more sustainable systems. This requires shifts to higher value production, or higher yields with more inputs per unit of production and higher standards of management (more knowledge intensive). However, sustainable agriculture has to work within the bounds of nature not against them. Many yield improvements can be achieved by optimizing efficiency of external inputs rather than trying to maximize yields.
- The importance of off-farm income should not be underestimated because it: i) supplements cash flow on the farm, ii) generates an investment environment for improved land management, and therefore iii) reduce production pressures on land.

Sharing responsibilities for sustainability:

- Farmers and land managers must expand their knowledge of sustainable technologies and implement improved procedures of land stewardship. The preferred option is not to tell the farmer what to do (command and control legislation), but to create an enabling environment through policy interventions where farmers are more free to make the right choice. A policy environment where farmers are more empowered, but also held accountable, for achieving the objectives of sustainable land management is essential. However, sustainable land management is the responsibility of all segments of society. Governments must ensure that their policies and programs do not create negative environmental impacts, and society needs to define requirements for land maintenance and develop a "social" discount rate for future land use options that encourages the most sustainable use.
- Concerns for sustainable land management go beyond agriculture to include the legitimate interests of other aspects of land stewardship, including wildlife, waterfowl and biodiversity management. There is increasing evidence that society is demanding that farmers become stewards of rural landscapes, and that agriculture become more than simply putting

food on the table. Many of society's environmental values may not represent economic gains for farmers, however, and farmers cannot shoulder all the costs of environmental maintenance.

Relationships among Soil Quality, Land Quality and Sustainable Land Management

New concepts of soil and land quality are emerging, and often these are used interchangeably. These concepts and their relationships are mentioned below, to the extent that some concensus is available on how these should be applied:

Soil quality is the capacity of a specific soil to function within natural or managed ecosystem boundaries to sustain plant and animal production, maintain or enhance water quality, and support human health and habitation.

Land quality is the condition, state or "health" of the land relative to human requirements, including agricultural production, forestry, conservation, and environmental management. Sustainable land management combines technologies, policies, and activities aimed at integrating socio-economic principles with environmental concerns so as to simultaneously maintain or enhance production, reduce the level of production risk, protect the potential of natural resources and prevent (buffer against) soil and water degradation, be economically viable, and be socially acceptable.

These concepts span the scales of detail, application, and levels of integration with socio-economic data. Soil quality is the most restrictive, followed by land quality and then sustainable land management. Soil quality is effectively a condition of a site, and it can be studied using soil data alone. Land quality requires integration of soil data with other biophysical information, such as climate, geology and land use. Land quality is a condition of the landscape, i.e. it is a biophysical property, but includes the impacts of human interventions (land use) on the landscape. Sustainable land management requires the integration of these biophysical conditions, i.e. land quality, with economic and social demands. It is an assessment of the impacts of human habitation, and a condition of sustainable development.

These are more than simple differences in semantics; the concepts differ in the kinds and scale of the processes being described, the data used for input, and the amount and kinds of integration with other disciplines. However, the concepts form a continuum over the landscape, and they must be applied for different types and scales of land use.

Land Quality Indicator Program

Assessment of sustainable land management requires appropriate evaluation instruments, such as Land Quality Indicators. However, land quality, like the concept of sustainable land management of which it is a part, requires operational definitions and specific, measureable indicators if it is to be more than an attractive, conceptual phrase.

The World Bank, in collaboration with UNEP, UNDP, FAO and the CGIAR, is developing a program called Land Quality Indicators (LQIs), as a means to better coordinate actions on land related issues such as land degradation. In the area of economic and social data, and in some cases for air and water quality, indicators are already in regular use to support decision-making at global, national and sub-national levels. In contrast, few such indicators are available to assess, monitor and evaluate changes in the quality of land resources. Land refers not just to soil but to the combined resources of terrain, water, soil and biotic resources that provide the basis for land use. Land quality refers to the condition or "health" of land, and specifically to its capacity for sustainable land use and environmental management.

The LQI program addresses the dual objectives of environmental monitoring as well as sector performance monitoring for managed ecosystems (agriculture, forestry, conservation and environmental management). It is being developed for application at national and regional scales, but it is also part of a larger, global effort on improved natural resources management. The LQI program is in response to the United Nations Conference on Environment and Development, and it fits with Agenda 21 expectations as well as the Convention to Combat Desertification.

Agricultural cropland, including agroforestry, as well as forested lands, range and pasture lands, are under increasing pressure because of population migrations to marginal land areas and agricultural intensification on existing cultivated lands. Sustainable land use intensification requires the maintenance or enhancement of the productive potential of the land resources, i.e. increases in food supplies must come from agricultural intensification rather than from area expansion, but this must be done without degrading the land resource on which production depends. The question is how this should be achieved, and how to monitor progress towards this objective in the different agroecological regions of developing countries. LQIs are tools to help us along the way.

In general and particularly in developing countries, it is essential that scarce resources devoted to land management be used more cost-efficiently and that policy-makers have at least rough indicators of whether environmental conditions and land quality are getting better or worse. Land quality indicators, such as nutrient balance, loss of organic matter, land use intensity and diversity, and land cover are useful to task managers and decision makers to monitor and improve the performance of projects with respect to their socio-economic and environmental impacts, and to assess the trend towards or away from land use sustainability. While routine project performance indicators based on cost-benefit analyses (input-output factors, risk and economic performance indicators), are necessary to monitor the activities and components of a project, LQIs are required to evaluate the environmental impact(s). The quantitative assessment of physical impacts, such as depletion of soil nutrients, loss of organic matter, soil erosion, water contamination etc. may appear to be costly and cumbersome during project implementation, but the long-term negative impact of reduced land quality, such as decreased efficiency of fertilizers, increased erosion, increased fuel consumption, increased pest infestation (nematodes, etc.), often result in rehabilitation costs that are much higher. The LQI approach focuses on preventive maintenance rather than rehabilitation, and provides the methodology and the approach to integrate the socio-economic and biophysical information that are required for better informed sustainable land management strategies.

The development of LQIs follows a logical framework, providing information not only on the state of the resources, but also the underlying causes (or "pressure") as well as the response of the society to the state and the pressure exerted on the land resources. Because of the nature and the complexity of land issues, the LQI program recommends addressing issues of land management by agro-ecological zones (Resource Management Domains or "Terroirs"). This new approach puts the focus on evaluating impacts of human interventions on specific landscapes, rather than emphasizing only biophysical variables as was the case in the past. At the same time, this spatial stratification favors incorporating farmer and other local knowledge into the overall process of improved agricultural and environmental land management.

A research strategy for the Land Quality Program was developed during a two day research planning meeting, sponsored by the World Bank. A panel of internationally acclaimed scientists and administrators established the objectives and priorities for the research, defined strategic alliances to be developed with ongoing national and international programs, and identified potential sources of funding. They also achieved international agreement on a core set of strategic land quality indicators.

Core LQIs for managed ecosystems (agriculture and forestry) in the major agro-ecological zones (AEZs) of tropical, sub-tropical and temperate environments, and recommended for development in the short term include:

- Nutrient balance: This describes nutrient stocks and flows as related to different land management systems used by farmers in specific AEZs and specific countries.
- Yield trends and yield gaps: This describes current yields, yield trends, and actual:potential, farm level yields for the major food crops in different countries.
- Land use intensity: Describes the impacts of agricultural intensification on land quality. Intensification may involve increased cropping, more value-added production, and increased amounts and frequency of inputs; emphasis is on the management practices adopted by farmers in the transition to intensification.
- Land use diversity (agro-diversity): Describes the degree of diversification of production systems over the landscape, including livestock and agro-forestry systems; it reflects the degree of flexibility (and resilience) of regional farming systems, and their capacity to absorb shocks and respond to opportunities.
- Land cover: Describes the extent, duration and timing of vegetative cover on the land during major erosive periods of the year. It is a surrogate for erosion, and along with land use intensity and diversity, it will give increased understanding on the issues of desertification.

A second set of core LQIs were recommended for longer-term research. These are indicators which require further development of their theoretical base, or lack adequate data for immediate development. These include:

- Soil Quality: Likely to be based on soil organic matter turn-over, particularly the dynamic (microbiological) carbon pool most affected by environmental conditions and land use change.
- Land degradation (erosion, salinization, compaction, organic matter loss): These processes have been much researched and have a strong scientific base, but reliable data on extent and impacts are often lacking.
- Agro-biodiversity: Involves objectives of managing natural habitats and the co-existence of native species in agricultural areas, maintaining natural soil micro and meso biodiversity, and managing the gene pools utilized in crop and animal production.

Four additional sets of core LQIs were identified, but these were recommended to be developed through collaboration with the respective authoritative disciplines:

- Water quality,
- Forest land quality,
- Rangeland quality,
- Land contamination/pollution.

The above are the biophysical components of sustainable land management. Although useful in their own right, they must still be complemented with indicators of the other pillars of sustainable land management, economic viability, system resilience, and social equity and acceptability. Considerable additional work is required to develop these pillars to the same level of detail as the land quality (biophysical) indicators.

2

Agroecosystem

Agroecosystem is an ecosystem on agricultural land that includes the diverse living and non-living components as well as their interactions. Some of the microbial impacts on agroecosystem include nutrient cycling and nitrogen fixation. The topics elaborated in this chapter will help in gaining a better perspective about agroecosystem as well as their management.

An agricultural ecosystem is an ecosystem managed with a purpose, usually to produce crops or animal products. Agricultural ecosystems are designed by humans, and are based on a long chain of experience and experiments. The emphasis in, for example, Western Europe has changed from maximum productivity only to also include environmental considerations, such as reduction of nutrient losses to groundwater and maintaining an open landscape with high biodiversity, etc. In less-productive regions such as sub-Saharan Africa, environmental considerations still have low priority.

Agricultural ecosystems comprise almost 40% (5 Gha) of the total land area of the Earth. About 11% of the total land area is arable land (cultivated with crops), and approximately 27% of the total land area is under permanent pasture, grazed by cattle, goats, sheep, camels, etc.

Plant biodiversity is extremely low – if weed control is successful there may be only one species present. In spite of this, belowground biodiversity can be high, although often lower than in natural ecosystems.

Ecological research performed in agricultural systems has advantages compared with research in most natural ecosystems, since agricultural fields are 'homogenized', that is, trees, larger stones, etc. are removed and regular soil cultivation to some extent evens out differences in topsoil properties over time.

Agroecosystems comprise 30% of the Earth's surface, and can be defined as, "the ecosystems in which humans have exerted a deliberate selectivity on the composition of the biota i.e., the crops and the livestock maintained by the farmer, replacing to a greater or lesser degree the natural flora and fauna of the site." Agroecosystems provide various ecosystem services, and management practices used in the agroecosystems determine the state of the global environment. However, most agroecosystems are disturbed more frequently and with greater intensity than natural ecosystems resulting in reduced biological diversity. Poor land management practices within many agroecosystems result in reduced soil organic carbon (SOC) and crop production, and increased erosion. About 50% of the global arable land is already under mechanical and chemical intensive agriculture, which requires high inputs of nutrient, energy, and water. Sustainable intensification often reduces mechanical and chemical inputs and increases biological diversification of agroecosystems. Examples include the use of diverse crop rotations, cover crops, no-tillage (NT), and the integration of livestock onto cropped lands. The adoption of diversification in agroecosystems has several advantages including building SOC, reducing insects/pests and diseases, and over time may result in improved crop productivity and other ecological services. However,"manipulating biotic interactions is not necessarily gentler than conventional agriculture and may also have undesirable effects." Sustainable intensification based on diversification of agroecosystems also includes a number of concerns such as higher initial cost, increased management skills, and an increase in uncertainty during the transition from mechanical and chemical intensification to sustainable intensification. Diversification in agroecosystems can be defined as "Diversity is at once a product, a measure, and foundation of system's complexity—and therefore, of its ability to support sustainable functioning. From this perspective, ecosystem diversity comes about as a result of ways that different living and nonliving components of the system are organized and interact. From another perspective, diversity as manifested by the complex of biogeochemical cycle and the variety of living organisms—is what makes the organization and interaction of the system possible." Diversification can generate greater employment opportunities and higher incomes for producers. A few examples of diversified farming systems include complex crop rotations, cover crops, and integrated crop-livestock (ICL) systems.

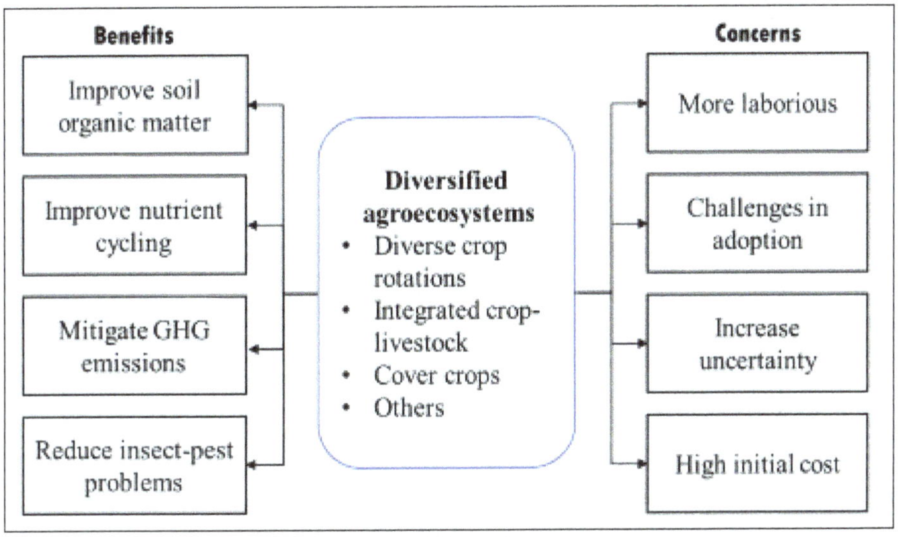

Benefits and concerns of diversified agroecosystems.

AGRICULTURAL BIODIVERSITY

Agricultural biodiversity is a broad term that includes all components of biological diversity of relevance to food and agriculture, and all components of biological diversity that constitute the agricultural ecosystems, also named agro-ecosystems: the variety and variability of animals, plants and micro-organisms, at the genetic, species and ecosystem levels, which are necessary to sustain key functions of the agro-ecosystem, its structure and processes.

Agricultural biodiversity is the outcome of the interactions among genetic resources, the environment and the management systems and practices used by farmers. This is the result of both natural selection and human inventive developed over millennia.

The following dimensions of agricultural biodiversity can be identified:

1. Genetic resources for food and agriculture:

   - Plant genetic resources, including crops, wild plants harvested and managed for food, trees on farms, pasture and rangeland species,
   - Animal genetic resources, including domesticated animals, wild animals hunted for food, wild and farmed fish and other aquatic organisms,
   - Microbial and fungal genetic resources.

   These constitute the main units of production in agriculture, and include cultivated and domesticated species, managed wild plants and animals, as well as wild relatives of cultivated and domesticated species.

2. Components of biodiversity that support ecosystem services upon which agriculture is based. These include a diverse range of organisms that contribute, at various scales to, *inter alia*, nutrient cycling, pest and disease regulation, pollination, pollution and sediment regulation, maintenance of the hydrological cycle, erosion control, and climate regulation and carbon sequestration.

3. Abiotic factors, such as local climatic and chemical factors and the physical structure and functioning of ecosystems, which have a determining effect on agricultural biodiversity.

4. Socio-economic and cultural dimensions: Agricultural biodiversity is largely shaped and maintained by human activities and management practices, and a large number of people depend on agricultural biodiversity for sustainable livelihoods. These dimensions include traditional and local knowledge of agricultural biodiversity, cultural factors and participatory processes, as well as tourism associated with agricultural landscapes.

Increases in production have traditionally involved increased dependence on pesticides and fertilizers and overuse of water that can degrade soils and water resources. Cultivation practices, however, are undergoing a shift to move from dependency on traditional inputs and from chemical-based intensification to forms of biological intensification that draws on the richness of biodiversity and natural resources to provide ecosystem services. Promoting the healthy functioning of ecosystem services, and doing so through ecosystem-based approaches, can increase the options available for optimizing agricultural production sustainably.

Biodiversity for Food and Agriculture

Biodiversity – the variety and variability of animals, plants and micro-organisms at the genetic, species and ecosystem levels – is necessary to sustain key functions of the ecosystem, its structure and processes. Biodiversity for food and agriculture can be managed to maintain or enhance ecosystem functions to provide options for the optimization of agricultural production, and contribute to the resilience of ecosystems for risk mitigation. Indeed, biodiversity enhances ecosystem services because those components that appear redundant at one point in time become important when changes occur.

Crop genetic diversity has a critical role to play in increasing and sustaining production levels and nutritional diversity throughout the full range of different agroecological conditions. Diverse organisms contributing to soil biodiversity perform a number of vital functions that regulate the soil ecosystem, including: decomposition of litter and cycling of nutrients; converting atmospheric nitrogen to an organic form, and reconverting this to gaseous nitrogen; and altering soil structure. Diversity in deliberate plantings on-farm through crop rotations, crop species mixtures, permanent soil cover crops employed in conservation agriculture or agroforestry – are oft used techniques to increase yield stability, and increase soil fertility. Grassland and pasture/crop systems that diversify and integrate ruminant livestock and crops tend to be more sustainable because they provide opportunities for rotation diversity, perennial cultivation, and greater energy efficiency.

The introduction of grazing animals at certain points in farming cycles may help to break down plant material and increase nutrient availability. Predators and parasites which attack pest insects or pathogens on crops, or plant-feeding insects which attack crop weeds contribute to pest regulation. Beyond these direct trophic relationships, a web-like pattern of interactions amongst diverse life-forms on-farm can deliver additional benefits. For instance, crop production may benefit from benign micro-organisms which colonize crops and their habitats such that pathogens do not establish, or from non-crop plants which are attractive to pests and thereby reduce their numbers on crops. Taken together, this directly- and indirectly-acting biodiversity may create "pest suppressive" conditions. Greater on-farm diversity of plants, greater closeness between crop plants and thus coverage of bare ground, and more perennial cultivation may be measures lending greater resistance to invasion of farming systems by noxious species, and assist in weed management. Pollinators are essential for orchard, horticultural and forage production, and contribute to improvements in quality of both fruit and fiber crops. Healthy pollination services are best ensured by an abundance and diversity of pollinators, in large part provided by wild biodiversity.

Ecosystem Services and their Importance for Agriculture

Ecosystem services are defined as "the benefits provided by ecosystems to humans". Many key ecosystem services provided by biodiversity, such as nutrient cycling, carbon sequestration, pest regulation and pollination, sustain agricultural productivity. Promoting the healthy functioning of ecosystems ensures the resilience of agriculture as it intensifies to meet growing demands for food production. Climate change and other stresses have the potential to make major impacts on key functions, such as pollination and pest regulation services. Learning to strengthen the ecosystem linkages that promote resilience and to mitigate the forces that impede the ability of agro-ecosystems to deliver goods and services remains an important challenge.

Ecosystem services can be:

- Supporting (e.g. soil formation, nutrient cycling, primary production).
- Provisioning (e.g. food, fresh water, fuelwood, fiber, biochemicals, genetic resources).
- Regulating (e.g. climate regulation, disease regulation, water regulation, water purification, pollination).
- Cultural (e.g. spiritual and religious, recreation and ecotourism, aesthetic, inspirational, educational, sense of place, cultural heritage).

Managing Biodiversity and Ecosystem Functions for Sustainable Agricultural Production

Biodiversity is an important regulator of agro-ecosystem functions, not only in the strictly biological sense of impact on production, but also in satisfying a variety of needs of the farmer and society at large. Agroecosystem managers, including farmers, can build upon, enhance and manage the essential ecosystem services provided by biodiversity in order to work towards sustainable agricultural production. This can be achieved though good farming practices which follow ecosystem-based approaches designed to improve sustainability of production systems. They aim at meeting consumer needs for products that are of high quality, safe and produced in an environmentally and socially responsible way.

The conservation and enhancement of biodiversity in cropping systems both above and below ground (e.g. soil biodiversity) are part of the foundation of sustainable farming practices. Such measures also lead to improved biodiversity in other parts of the environment which are adjacent to but not directly part of the cropland – such as water bodies and the broader agricultural landscape. The composition and diversity of planned biodiversity (for example selected crops) strongly influences the nature of the associated diversity – plant, animal and microbial. A challenge is to integrate, through ecosystem approach strategies, the planned biodiversity that is maintained with the associated diversity (for example wild pollinators).

Some Ecosystem-based Production Practices

- Conservation Agriculture (CA),
- Integrated rice management systems such as the System of Rice Intensification (SRI),
- Integrated Pest Management (IPM) for plant and animal diseases,
- Integrated Production and Pest Management (IPPM),
- Integrated Plant Nutrition Systems (IPNS),
- Integrated Weed Management (IWM).

Maintenance of biodiversity, including crop-associated biodiversity, is necessary to ensure the continued supply of goods and services such as:

1. Evolution and crop improvement through breeding: The interaction between the environment, genetic resources and management practices that occurs in situ within agro-ecosystems ensures that a dynamic portfolio of biodiversity for food and agriculture (agricultural biodiversity) is maintained and adapts to changing conditions.
2. Biological support to production: Support is provided by the organisms that make up the biological diversity of the agro-ecosystem. For example, soil fauna and micro-organisms, together with the roots of plants and trees, ensure nutrient cycling; pests and diseases are kept in check by predators and disease control organisms, as well as genetic resistances in crop plants themselves; and insect pollinators contribute to the cross-fertilization of outcrossing crop plants.

3. Wider ecological functions: Valuable ecological processes that result from the interactions between species and between species and the environment include the maintenance of soil fertility, water quality and climate regulation.

Microorganisms in Soil

Microorganisms are very small forms of life that can sometimes live as single cells, although many also form colonies of cells. A microscope is usually needed to see individual cells of these organisms. Many more microorganisms exist in topsoil, where food sources are plentiful, than in subsoil. They are especially abundant in the area immediately next to plant roots (called the rhizosphere), where sloughed-off cells and chemicals released by roots provide ready food sources. These organisms are primary decomposers of organic matter, but they do other things, such as provide nitrogen through fixation to help growing plants, detoxify harmful chemicals (toxins), suppress disease organisms, and produce products that might stimulate plant growth. Soil microorganisms have had another direct importance for humans—they are the source of most of the antibiotic medicines we use to fight diseases.

Bacteria

Bacteria live in almost any habitat. They are found inside the digestive system of animals, in the ocean and fresh water, in compost piles (even at temperatures over 130 °F), and in soils. Although some kinds of bacteria live in flooded soils without oxygen, most require well-aerated soils. In general, bacteria tend to do better in neutral pH soils than in acid soils.

In addition to being among the first organisms to begin decomposing residues in the soil, bacteria benefit plants by increasing nutrient availability. For example, many bacteria dissolve phosphorus, making it more available for plants to use. Bacteria are also very helpful in providing nitrogen to plants, which they need in large amounts but is often deficient in agricultural soils. You may wonder how soils can be deficient in nitrogen when we are surrounded by it—78% of the air we breathe is composed of nitrogen gas. Yet plants as well as animals face a dilemma similar to that of the Ancient Mariner, who was adrift at sea without fresh water: "Water, water, everywhere nor any drop to drink." Unfortunately, neither animals nor plants can use nitrogen gas (N_2) for their nutrition. However, some types of bacteria are able to take nitrogen gas from the atmosphere and convert it into a form that plants can use to make amino acids and proteins. This conversion process is known as nitrogen fixation.

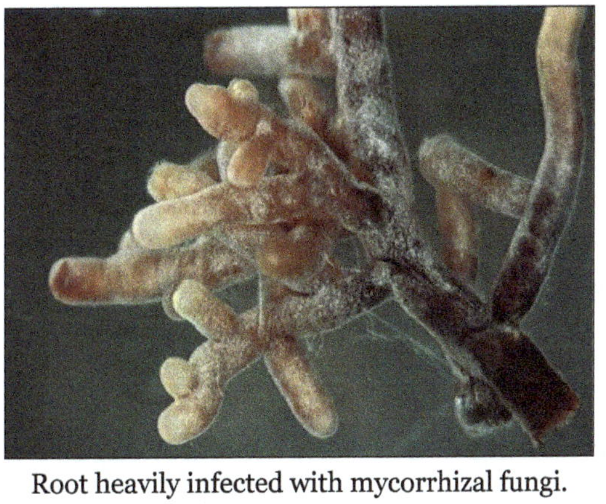

Root heavily infected with mycorrhizal fungi.

Some nitrogen-fixing bacteria form mutually beneficial associations with plants. One such symbiotic relationship that is very important to agriculture involves the nitrogen-fixing rhizobia group of bacteria that live inside nodules formed on the roots of legumes. These bacteria provide nitrogen in a form that leguminous plants can use, while the legume provides the bacteria with sugars for energy.

People eat some legumes or their products, such as peas, dry beans, and tofu made from soybeans. Soybeans, alfalfa, and clover are used for animal feed. Clovers and hairy vetch are grown as cover crops to enrich the soil with organic matter, as well as nitrogen, for the following crop. In an alfalfa field, the bacteria may fix hundreds of pounds of nitrogen per acre each year. With peas, the amount of nitrogen fixed is much lower, around 30 to 50 pounds per acre.

The actinomycetes, another group of bacteria, break large lignin molecules into smaller sizes. Lignin is a large and complex molecule found in plant tissue, especially stems, that is difficult for most organisms to break down. Lignin also frequently protects other molecules like cellulose from decomposition. Actinomycetes have some characteristics similar to those of fungi, but they are sometimes grouped by themselves and given equal billing with bacteria and fungi.

Fungi

Fungi are another type of soil microorganism. Yeast is a fungus used in baking and in the production of alcohol. Other fungi produce a number of antibiotics. We have all probably let a loaf of bread sit around too long only to find fungus growing on it. We have seen or eaten mushrooms, the fruiting structures of some fungi. Farmers know that fungi cause many plant diseases, such as downy mildew, damping-off, various types of root rot, and apple scab. Fungi also initiate the decomposition of fresh organic residues. They help get things going by softening organic debris and making it easier for other organisms to join in the decomposition process. Fungi are also the main decomposers of lignin and are less sensitive to acid soil conditions than bacteria. None are able to function without oxygen. Low soil disturbance resulting from reduced tillage systems tends to promote organic residue accumulation at and near the surface. This tends to promote fungal growth, as happens in many natural undisturbed ecosystems.

Many plants develop a beneficial relationship with fungi that increases the contact of roots with the soil. Fungi infect the roots and send out root-like structures called hyphae. The hyphae of these mycorrhizal fungi take up water and nutrients that can then feed the plant. The hyphae are very thin, about 1/60 the diameter of a plant root, and are able to exploit the water and nutrients in small spaces in the soil that might be inaccessible to roots. This is especially important for phosphorus nutrition of plants in low-phosphorus soils. The hyphae help the plant absorb water and nutrients, and in return the fungi receive energy in the form of sugars, which the plant produces in its leaves and sends down to the roots. This symbiotic interdependency between fungi and roots is called a mycorrhizal relationship. All things considered, it's a pretty good deal for both the plant and the fungus. The hyphae of these fungi help develop and stabilize larger soil aggregates by secreting a sticky gel that glues mineral and organic particles together.

Algae

Algae, like crop plants, convert sunlight into complex molecules like sugars, which they can use for energy and to help build other molecules they need. Algae are found in abundance in the flooded soils of swamps and rice paddies, and they can be found on the surface of poorly drained soils and in wet depressions. Algae may also occur in relatively dry soils, and they form mutually beneficial relationships with other organisms. Lichens found on rocks are an association between a fungus and an alga.

Protozoa

Protozoa are single-celled animals that use a variety of means to move about in the soil. Like bacteria and many fungi, they can be seen only with the help of a microscope. They are mainly secondary consumers of organic materials, feeding on bacteria, fungi, other protozoa, and organic molecules dissolved in the soil water. Protozoa—through their grazing on nitrogen-rich organisms and excreting wastes—are believed to be responsible for mineralizing (releasing nutrients from organic molecules) much of the nitrogen in agricultural soils.

ECOLOGICAL RESTORATION OF FARMLAND

The goal of ecological restoration is to shift an ecosystem towards its pre-disturbed state with respect to ecosystem structure, function and composition. The approach emphasizes the use of quantitative practices for measuring and restoring ecosystem 'health', including its ability to deliver ecosystem services. Sustainable agriculture 'refers to the ability of a farm to continue producing indefinitely with a minimum of outside inputs' or put another way, 'is defined as agriculture that meets the needs of the present generation while conserving resources for the use of future generations.' The continuity of production by using minimal inputs and creating few negative effects is emphasized. 'Farmland' primarily refers to the land use comprising temporary or permanent crops and pastures.

Why is Ecological Restoration of Farmland Necessary?

What causes farmland to become degraded and what are the symptoms of farmland in need of ecological restoration? Farmland and its environs are susceptible to inadvertent or deliberate degradation in their physical, chemical and biological condition by a range of farming activities that primarily result in changes to air quality, biological diversity, climate, soil condition and the quality and quantity of water. Soil erosion results from the loss of vegetation cover due to burning, grazing and cultivation. Changes in the fertility, structure, acidification and salinization of soils are caused by cultivation, drainage, irrigation and tree removal. Pollution of ground water and eutrophication of rivers and lakes results from off-farm movement of silt, pesticides and nutrients, i.e. fertilizers or animal effluent. Flow rate of rivers is affected by the construction of weirs and levée-banks, diversion of overland water flows to on-farm reservoirs and direct removal for irrigation. There are global impacts on atmospheric constituents (principally carbon dioxide, methane and nitrogen dioxide) and climate (chiefly temperature and rainfall) as a result of forest removal, biomass burning, fertilizers and livestock. Finally, land cover changes lead to both habitat loss and fragmentation, which threaten aquatic and terrestrial taxa.

Symptoms of degraded farmland include algal blooms and pesticide residues in waterways, pest outbreaks, plant disease epidemics such as 'rural dieback' of native Australian eucalypts, which is principally caused by the root rot fungus *Phytophthora cinnamomi*, and disease epidemics of livestock, such as foot and mouth disease and influenza A virus (H5N1, 'bird flu'). In addition, there is evidence of yield decline, loss of topsoil through water and wind, hedgerows and field margins removed or sprayed with herbicides, and a general reduction in species richness and abundance of plants and animals.

Importantly, agricultural practices have both local and landscape-scale impacts that transcend farm boundaries. Local intensification includes adverse effects such as shortened crop rotation cycles and increasing input of agrochemicals. On a landscape scale, fields have been amalgamated and enlarged, resulting in simplified landscapes with few or no non-crop habitats remaining. The total annual external (off-farm) costs of agriculture on natural resources (air, soil and water), biodiversity and human health (pathogens and pesticides) have been estimated for the United Kingdom at £1149–3907 million between 1990 and 1996 and £1514 million in 2000, and for the USA at £3256–9678 million in 2002. This equates to £208 ha^{-1} of arable land and permanent pasture for the United Kingdom (UK) in 1996 and £17–55 ha^{-1} of arable land in the USA.

To put these costs into perspective with the external benefits provided by agriculture, the pivotal paper by calculated the combined economic value of three ecosystem services (biological control, pollination and food production) from *worldwide* cropland to be US $128.8 (£73.802) billion per year or US $92 (£53) per ha. A caveat here is that assigned a nil value to the ecosystem service of habitats or refugia for resident and transient taxa because this service 'does not occur or is known to be negligible', hence the true value of cropland is likely to be underestimated. Nevertheless, by these calculations, the worldwide ecosystem service benefits from agriculture are estimated to be £53$ha^{-1}yr^{-1}$, yet the external costs of intensive agriculture in countries like the UK are £208$ha^{-1}yr^{-1}$. Equally compelling calculations estimated that the economic benefit to world society from biodiversity is US $2928 billion. This value included the benefits of activities such as biological pest control, ecotourism, pollination and waste disposal. It is evident that more sustainable agricultural practices in conjunction with ecological restoration methods on farmland are necessary to reduce the unacceptably high external costs of agriculture that are borne by the community. In addition, 'ecological engineering' techniques are available to enhance ecosystem services on farmland, including habitat manipulation tactics for beneficial arthropods that are responsible for biological pest control and contribute to biodiversity in general.

Agriculture and biodiversity conservation have been traditionally viewed as incompatible, with agriculture considered a major driver of species loss for many plant and animal taxa, such as bumble-bees (*Bombus* spp.) and bird species like skylarks (*Alauda arvensis* L.) since 1945. Agriculture represents the dominant land use throughout much of western Europe and a significant part of European biodiversity is associated with this habitat. Agroecosystems, however, are very hostile to a wide diversity of species owing to the conversion of complex natural ecosystems to simplified managed ones and the intensification of resource use. Firstly, there is a tendency for simplified cropping systems to be applied to increasingly consolidated land areas, leading to the loss of non-crop habitats, such as field margins and hedgerows, together with the decline in traditional mixed arable and livestock farming. As a result, remnant native vegetation has become fragmented into different patches and there are fewer 'nodes' where field corners join. These nodes can be rich 'hotspots' of invertebrate, vertebrate and plant diversity). Secondly, there is intensification of resource use in the cropping systems themselves, including greater pesticide and fertilizer usage and shorter fallow periods. However, more recently, there has been an important move beyond conservation efforts to an appreciation of the value of natural, undisturbed remnants and to a better recognition of the role that highly modified landscapes play in maintaining native biodiversity. As pointed out 'we are obviously past any point where strategies that focus on preservation of 'pristine' habitats are sufficient for the job. Greater attention must be placed on human-dominated landscapes that (surround) the less disrupted areas'. In this way, agriculture can make important contributions to high-diversity habitats, while also benefiting from ecosystem services provided from different land use types. We know that invertebrate natural enemies of crop pests visit different habitat types before colonizing agricultural fields and improved biological pest control and crop pollination may directly increase farmers' income.

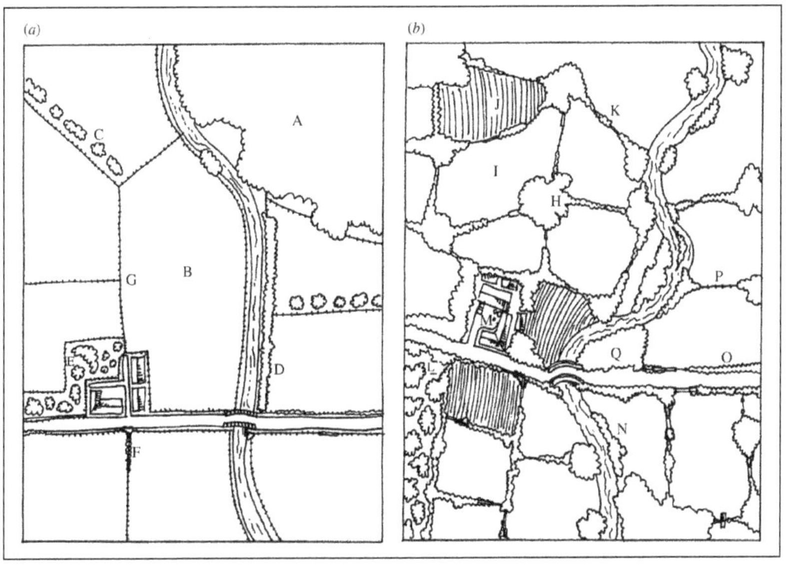

Contrast between (*a*) a simple (or homogeneous) and (*b*) a complex (or heterogeneous) farm landscape. Note the heterogeneity of habitats and the connectance features in the complex landscape compared with the uniformity of the simple farm landscape. A) indigenous plant reserve: these tend to be large tracts of land not integrated with farmland; B) pasture and exotic grasses; C) typical shelter belt (e.g. popular, *Cupressus macrocarpa* and *Pinus radiata*); D) riparian vegetation (e.g. willow, grasses and some indigenous species); E) farmhouse garden; F) small areas of patchy gorse, *Ulex europaeus*; G) wire fences: common field boundaries; H) small wood lot: a highly used but sustained feature; I) pasture; J) ploughed field; K) hedge fence; L) orchard; M) farmhouse garden; N) riparian vegetation; O) roadside vegetation, hedges, trees, etc; P) wire fences or stone walls; and Q) woodland.

While the rate of conversion of land to agriculture has slowed in more developed countries like the United Kingdom, ecosystem function in fragmented remnant vegetation often remains severely disturbed due to continuing effects of microclimatic change and isolation of taxa pointed out that populations that are too small to be viable might still persist for long periods, simply owing to the longevity of remaining individuals. For example, remnant native areas of the Western Australian wheat belt, which now represent only 7% of 14 million ha, contain female trapdoor spiders *Anidiops villosus* (Rainbow) that can live for over 23 years (Thus, the presence alone of a species in a remnant is no guarantee of its continued existence there, as it also requires successful reproduction and recruitment, the success of which can be revealed by a closer examination of the species' age structure. A key objective in conservation and restoration ecology is the determination of the minimum viable population sizes for a given habitat size that can sustain a given community of plants and animals, as well as the metapopulation effects of fragment shape and position in the landscape. Vegetation 'corridors' that link otherwise isolated patches have been suggested for improving the persistence of taxa, and there is a general consensus that they are beneficial. Thus, agroecosystems do support biodiversity that contributes overall to ecosystem services such as crop pollination and biological pest control, and ecological restoration can to some extent ameliorate the effects of agricultural expansion and intensification. Finally, while farmland ecological restoration methods may serve their immediate purpose of contributing to sustainable agriculture, other non-restoration or 'multi-function' benefits may be achieved. For example, the advantages of planting trees on farmland include: bioenergy production, carbon sequestration, erosion control, habitat restoration, increased water use (reduced secondary salinity) and wood production (farm forestry), as well as aesthetics, cultural, landscape conservation and other benefits.

Target Species of Farmland Ecological Restoration

There has been considerable debate in the ecological restoration literature about whether the requirements of single species should be the basis of designing conservation programmes, or whether the analysis of landscape attributes that support entire communities should underpin conservation planning. In a single-species approach, rare or vulnerable species or groups of species that are considered to represent important components of biodiversity are often the targets. But it has been argued that the single-species approach can be costly and ineffective in dealing with the urgency of threats to ecosystems and their functioning. Conversely, it is considered unwise to ignore the requirements of individual species when seeking to define the attributes of an improved landscape that will ensure community survival via enhancement of species richness and evenness. The concepts of 'flagships', 'umbrellas', 'biodiversity indicators', 'focal-species' and other classes of taxon-based restoration schemes have been developed by the proponents of the single-species approach as an alternative to studying each species separately. In focal-species schemes, the species are purportedly selected on the basis that their requirements encapsulate those of the regional taxa. In practice, the species are either classified as being area sensitive, dispersal limited, resource limited or limited by ecological processes such as fire and predation. However, critics have challenged the underlying validity of the assumption that such taxa can be adequately chosen to represent the regional taxa. For instance, examined the outcome of how well threatened, rare or 'of concern' species that were deliberately selected for inclusion in at least 14 different surrogate schemes (e.g. large carnivores, charismatic species, habitat specialists) actually do represent the interests of the total pool of species across all schemes. Also, the area of habitat that would be protected compared with conserving suites of species that were randomly selected from each of three conservation databases is a concern. The results showed that the various surrogate schemes were generally poor at protecting the total background pool of taxa. As an alternative to a focal-species approach alone, urged the adoption of a mix of approaches across different landscapes, leading to economically, politically and socially acceptable results. This form of a bet-hedging strategy would help in the event that a single approach was unsuccessful in a given location. Examples include habitat restoration to a nominal proportion of vegetation cover, expansion of the size of existing patches of remnant vegetation, and restoration of a watercourse and associated riparian vegetation. urged that the response of a large set of species within ecosystems should be measured in response to ecological restoration changes. They also highlighted the need for continued basic autecological research, although data intensive, to provide the foundations for ecological restoration practices, such as: (i) abundance and distribution changes, (ii) dispersal ability, (iii) the influence of adjoining land use types on behaviour, (iv) the spatial scale at which organisms respond to disturbances such as fire and grazing, and (v) the cumulative effects of disturbance. Further, it was recognized that it is better to explain openly than to hide the complexities and uncertainties involved in ecological restoration and to desist from the promotion of 'simple rules' for restoration to landholders, like the focal-species approach. The complexities of species-diverse food webs are well illustrated by, who considered the concepts of species identity (i.e. the 'sampling' effect) and niche complementarity when managing biodiversity to enhance pest suppression in farmland.

Despite the above cautions, it is insightful to consider the taxa that have been the subject of single- or focal-species farmland ecological restoration approaches. Often the target species are classified as rare or threatened (at risk of extinction) based on their population size and trend, like the skylark (A. arvensis) in Europe. Host or prey species that are important in the diet of the species under conservation are also targeted, such as various insects. Conversely, predator species that threaten the existence of species under conservation are targeted, such as the red fox (Vulpes vulpes L.) and ferret (Mustela furo L.). Species may be considered good indicators of habitat quality, such as butterflies and insectivorous bats (e.g. Pipistrellus pipistrellus (Schreber)). Other species may be common and fulfil important ecosystem services, such as biological control, nutrient cycling and pollination. Examples include species of carabid beetles and bumble-bees. Finally, attempts are made to reduce populations of non-native or native species that cause disruption to agro-ecosystem function in a given locale; examples are a freshwater fish the common carp (Cyprinus carpio L.) and lantana weed (Lantana camara L.). It is recognized that species that fall into the above category could also be classed as agricultural or invasive pests, which are categories not represented here, but the intention is to demarcate targets of agricultural pest management and biosecurity eradication programmes within farmland ecological restoration. An evaluation of agri-environmental schemes in Europe recorded the response of birds in 29 published studies, plants in 20, insects and spiders in 20 and mammals in only one.

Table: Examples of plant and animal taxa that have been targeted in farmland ecological restoration programmes.

Rationale	Target group	Species	Country
Threatened or rare	Farm birds	Corn bunting (Miliaria calandra); skylark (Alauda arvensis); yellowhammer (Emberiza citronella)	United Kingdom
	Game birds	Grey partridge (Perdix perdix); pheasant (Phasianus colchicus); red-legged partridge (Alectoris rufa)	France
	Raptors	Hen harrier (Circus cyaneus); NZ falcon (Falco novae-seelandiae)	United Kingdom, New Zealand
	Mammals	Field vole (Microtus agrestis); harvest mouse (Micromys minutus)	United Kingdom
	Insects and other invertebrates	Bumble-bees (Bombus spp., Hymenoptera); butterflies (Lepidoptera); wetas (Orthoptera)	Australia, New Zealand, United Kingdom
	Plants	Kanuka (Kunzea ericoides); kowhai (Sophora microphylla); totara (Podocarpus totara)	New Zealand

Prey ('chick-food')	Insects and other invertebrates	Beetles (Coleoptera); butterflies (Lepidoptera); grasshoppers (Orthoptera); harvestmen (Opiliones); spiders (Araneae); true bugs (Hemiptera)	United Kingdom
Predators of threatened species	Mammals	American mink (Mustela vison); Australian brushtail possum (Trichosurus vulpecula)a; badger (Meles meles); ferret (Mustela furo)a; hedgehog (Erinaceus europaeus); weasel (Mustela nivalis); red fox (Vulpes vulpes); stoat (Mustela erminea)	France, New Zealand, United Kingdom
Indicator of habitat quality	Birds	Port Lincoln ringneck (Platycercus zonarius); singing honeyeater (Lichenostomus virescens); yellow-throated miner (Manorina flavigula)	Australia
	Insects and other invertebrates	Beetles (Coleoptera); butterflies (Lepidoptera); flatworms (Turbellaria); spiders (Araneae)	Finland, New Zealand, United Kingdom
	Mammals	Bats (Pipistrellus pipistrellus)	United Kingdom
	Vascular plants	Heather (Calluna vulgaris)	Denmark
Biological control agents	Insects and other invertebrates	Beetles (Coleoptera: Carabidae, Staphylinidae); spiders (Araneae); wasps (Hymenoptera)	Australia, New Zealand, United Kingdom, USA
Decomposers	Insects and other invertebrates	Dung beetles (Coleoptera: Scarabaeidae)	Ireland
Pollinators	Insects and other invertebrates	Bumble-bees (Bombus spp., Hymenoptera); hoverflies (Diptera)	New Zealand, United Kingdom
Damaging (non-native	Birds	Noisy miner (Manorina melanocephala)	Australia
or native)	Fishes	Common carp (Cyprinus carpio)	Australia
	Insects and other invertebrates	Honeybee mite (Varroa destructor, Acari: Varroidae)	New Zealand
	Mammals	European rabbit (Oryctolagus cuniculus)	New Zealand
	Plants	Lantana (Lantana camara)	Australia

Also regarded as pest species because they vector bovine tuberculosis to cattle and deer in New Zealand.

A salient consideration in ecological restoration is the concepts of multitrophic interactions and polyphagy. The work of is a recent example of how 'beetle banks' (raised strips of unsprayed and uncultivated perennial grasses) increased predatory beetle (Coleoptera: Carabidae and Staphylinidae) abundance adjacent to the banks, but not rates of predation on fly eggs due to intraguild predation and the presence of alternative prey in this area. The presence of the noisy miner (Manorina melanocephala Latham), an aggressive bird species native to Australia, can affect avian diversity and abundance in remnant patches of woodland on Australian farmland. Moreover, their presence is believed to exclude small insectivorous birds that play an important role in reducing the abundance of leaf-feeding insects that contribute to 'rural dieback' of native trees like eucalypts.

There seems to be no worldwide compendium of species that have been targeted in farmland ecological restoration programmes. It probably reflects the wide range of habitat types that farmland as a land use type occupies. There are, however, various publications that list the conservation status of individual species present by biome, habitat or ecoregion, such as grassland, savannah and shrubland, which would represent farmland. The most notable publication is the IUCN World Conservation Union's Red List of Threatened Species, which presently recognizes 15589 species that are seriously threatened with extinction of the 38047 assessed species (i.e. 41% threatened). The seriously threatened species comprise 5274 species of vertebrates, 1992 invertebrates, 8321 vascular plants and two lichens from a range of countries across the world. However, it is claimed that this figure is an underestimate of the total number of threatened species as it is based only on an assessment of less than 3% of the world's 1545594 described species. This key publication has inspired numerous separate listings for various countries. For example, the population status of 247 bird species has been assessed in the United Kingdom; 40 species are critically threatened ('red-listed'), 121 are moderately threatened ('amber-listed') and 86 are rare or not presently threatened ('green-listed'). An action plan for Australian butterflies also exists, which is aimed at conserving threatened butterfly species and boosting their numbers. The action plan reviewed 220 of Australia's 654 butterfly species and found that 26 were threatened. For each species reviewed, there is a brief commentary on its distribution, specific habitat requirements, major threatening processes, historic and present conservation status and a recovery action plan. An unfortunate common feature of the above lists of threatened organisms is the recognized inability to formulate the conservation status and recovery plans for a large number of species either because few have been assessed or owing to a deficiency of data. Further, in many cases, it is not clear what the definitive cause of population decline was, although a wide variety of threats are generally implicated. The loss of habitat from agricultural expansion is generally the main reason suggested, but the evidence can be weak if it is based on anecdotal observations, unpublished data or 'grey' literature such as annual reports rather than empirical evidence in refereed scientific publications. Nevertheless, these lists are regarded as an extremely valuable tool for conservation planning, management, monitoring and decision making. However, what is the relevance of these lists for farmland ecological restoration? Key values are that they provide a framework to help identify species of conservation importance, suggest causes for decline and propose a recovery plan. In this context, it would be very useful if they explicitly listed which species also colonize farmland within a recognized habitat type, like native grassland, but they appear not to do this.

Empirical research has been used to support ecological restoration programmes by determining which species are more or less suitable for restoration. For instance, large-seeded and shade-tolerant native tree species were recommend rather than small-seeded and shade-sensitive species for reforestation of abandoned farmland in Panama. Intrinsic biological traits for a range of taxa have been identified as good correlates of susceptibility to local extinction. Depending on the taxa, susceptible traits include low abundance, high habitat specificity, large body size and slow reproductive rates. Interplay exists between these traits with external threats; for example, habitat loss tends to affect mostly those species that are ecologically specialized, while introduced predators may have more of an impact on species with long generation times. The relative contribution of external threats and intrinsic traits to extinction risk has been investigated for some taxa; among mammals approximately 50% of the variation in extinction risk is explained by variation in species' biological traits, with the remainder being attributable to human pressures and the interactions between these and biological traits.

When is Ecological Restoration Conducted?

It is intuitive that ecological restoration of degraded farmland should be initiated as soon as it is feasible and subsequent monitoring may last an indefinite period. In practice, however, monitoring rarely lasts for more than 5 years, even though this may be too short a period. The specifics of determining how long the ecological restoration takes in terms of reaching an 'endpoint' and how frequently monitoring should be conducted is highly site-specific and rather controversial, partly because it is hard to determine. This issue will be discussed further in 7. Nevertheless, the ecological succession of plant and animal communities has often been the focus of several farmland restoration assessments. One of the best examples is provided by, who examined the succession of vascular plant communities from farmland to heathland over a period of 22 years after cultivation ceased. The succession of plants and arthropods has also been studied in beetle banks over 7 years. The grasses Agrostis stolonifera L., Dactylis glomerata L. and Holcus lanatus L., but not Lolium perenne L., became well established when sown in single-species plots, and in mixtures L. perenne was completely replaced with D. glomerata by the third year. The abundance of predatory carabid and staphylinid beetles (Coleoptera) and linyphiid and lycosid spiders (Araneae) fluctuated over the seven consecutive winters; densities tended to be higher in the second winter than in the first, then fell in the third and fourth winters, followed by rises in the fifth and sixth and fell again in the seventh. A positive relationship was detected between arthropod diversity, but not density, and the age of the beetle bank. This indicated that arthropods respond to increasingly complex plant communities that develop in the beetle banks (there was a positive relationship between plant diversity and arthropod diversity both in beetle banks and field margins. In contrast, in another study, there was no effect of habitat age on the overall abundance or diversity of Araneae and Coleoptera in 'newly' sown tussock grass (3–4 years old) and mature field margins (ca 50 years old), or in newly planted (2–5 years old) and mature hedge bases (40–60 years old). This highlights the fact that many farmland arthropods have a good dispersal ability, which permits them readily to colonize different habitats regardless of age.

However, a large proportion of ecological restoration studies have been conducted only on newly created habitats in the early stages of succession (less than 3–5 years). In some cases, only very short-term changes of the order of days to a few weeks are of prime interest, such as in studies of movement and predation, which support the theoretical foundations of ecological restoration. Conversely, examined only the later stages of succession. In some cases, the age of the habitat being sampled is not stated and is possibly unknown.

The importance of the particular time of year in relation to seasonal lifecycles has also been the focus of several studies. Beetle banks provide essential overwintering habitat for many species of arthropods and birds. Natural and artificial nesting sites and supplementary food are reportedly helpful for breeding birds in the spring, as the density of arthropods (as chick food) tends to be lower in the spring than summer months in beetle banks and grassy field margins. While examined only the value of different habitats in providing forage (pollen and nectar) for bumble-bee populations in late summer, they urged that the food requirement in spring and early summer, together with suitable nesting and hibernation habitat, be considered for effective bumble-bee conservation. Similarly, though another study lasted only four weeks during the main period of bumble-bee forage and nest growth in early summer, it was suggested that the nest establishment phase in early spring when the queen bee has to gather sufficient forage to provision her first batch of offspring is a critical time when differences in availability of floral resources between habitats is most vital. Finally, it may be possible to minimize the effects of adverse husbandry practices, such as soil cultivation, planting and harvesting, on beneficial arthropods and other animals under conservation by uncoupling their temporal (and spatial) synchrony at the farm scale. For instance, spring-sown cereals provide better quality habitat for farmland birds than do autumn-sown cereals (e.g. winter wheat).

Where does Ecological Restoration Take Place?

According to the IUCN Red List of Threatened Species, there are more threatened species in terrestrial ecosystems (4427 species threatened of the 21053 assessed; 21%) than freshwater (1388 of the 5574; 25%) or marine (187 species threatened of the 843 assessed; 22%) worldwide (species that live in more than one ecosystem are counted more than once). Although the proportion of threatened species appears to be little different between ecosystems, this is thought to be an artefact of fewer marine and freshwater assessments; preliminary indications suggest that freshwater species may actually be relatively more seriously threatened than the terrestrial species. The absolute numbers of threatened species are unevenly distributed throughout the eight recognized terrestrial biogeographic realms of the world, with most species (including endemics) occurring in tropical areas in countries such as Australia, Brazil, China, Indonesia and Mexico. As mentioned earlier, the threatened species that also colonize farmland are not explicitly listed in this or any other 'red-list'. Nevertheless, a review of 68 restoration studies published in *Restoration Ecology* between 1993 and 2003 reported that the majority of studies were carried out in North America (53%), but that there was also relevant work in Australia (19%), Europe (16%), Africa (4%), South America (4%) and Asia (3%) ithin Europe, just six countries had contributed a total of 62 relevant evaluation studies between 1994 and 2000. These countries were: the United Kingdom (n=29 studies), the Netherlands (n=18), Germany (n=6), Switzerland (n=5), Ireland (n=3) and Portugal (n=1).

In terms of habitats most frequently examined, wetlands were most frequently studied (19%), followed by grassland (16%) and montane forest (13%). Further, ecological restoration was most commonly conducted at sites wherein the previous land uses were mining (37%), agriculture (18%; presumably arable cropping) and pasture. Recent studies on the response of different animals to various habitat types on farmland have considered taxa such as bats, farmland birds and butterflies.

Regional effects on biodiversity are also likely to occur within a given country. tested the hypothesis that the abundance and diversity of bumble-bees are greater in the more enclosed, mixed farming region of the West Midlands, compared with the open, intensive arable region of East Anglia in the United Kingdom. Despite the significantly higher abundance and plant species richness of dicotyledonous flowers on the lighter soils of the West Midlands, there were no significant differences in the abundance or richness of bumble-bee species recorded on the field margins between the two study areas. In Germany, the species richness of spiders (Araneae) was no different between the southern Lower Saxony region that had moderate to high amounts of arable land (25–85%) and the Central Hesse region, which had less arable land (7–61%). Indeed, a regional scale approach is presently being used to 'green' or restore ecologically the entire Waipara vineyard region of New Zealand. Flowering herbs and native shrubs and trees have been planted both within and surrounding the vineyards to restore components of previous ecosystems in a productive landscape and provide habitat (shelter and food resources) for native birds and beneficial arthropods. Critically, this conservation effort brings tangible tourism and marketing benefits. A typical rear label on a bottle from the Canterbury House vineyard in that region is dominated by 'added value' restoration words: 'This wine comes from one of the vineyards taking part in the Greening Waipara project'. Native vegetation is being restored for butterfly conservation and other benefits.

Several other studies have urged the development of effective restoration practices at a regional spatial scale. Indeed, and were highly critical of efforts to manage a certain habitat or a group of habitats within an area at 'human' scales that may in fact be inappropriate for the majority of the species present and the associated ecological processes. In this way, the 'ecological scales' to which various taxa respond and at which key processes occur should be adopted rather than human-perceived catchment or other scales. Hence, just as the negative effects of agricultural intensification can operate at different scales that transcend farm boundaries, it is becoming apparent that animals such as bumble-bees and spiders may respond to spatial and temporal changes in resource supply at scales greater than that of a single farm, and that habitat restoration measures may need to be targeted at the regional, rather than the local level. Intensive studies of landscape wide patterns of distribution and the tracking of animal movement are helping foster this appreciation of ecological conservation and restoration purposes. Some of this work has specifically examined the movement of animals between patches of arable land and non-crop areas like remnant vegetation.

Within a farming unit, the sites that have been the targets of ecological restoration include arable land for fallow or undersowing, areas adjacent to sensitive urban areas such as schools and roads, farm woodland, field margins including uncultivated areas (set aside), headlands and hedges, grassland, land surrounding aquifers and land water catchments, mid-field strips (beetle banks and cover strips), roadside verges, scrub, waterways and riparian areas. The paper by is a good example of how land attributes such as aspect, elevation, soil depth and slope can be used in farm-scale site selection for farm forestry.

How is Ecological Restoration Practised?

The inherent complexities of ecological restoration and idiosyncrasies of each project means it is unwise to duplicate the same efforts elsewhere, though the principles should be transferable. Nevertheless, we present the generic principles of ecological restoration and draw on specific examples. The Society for Ecological Restoration International (SERI) produced a document that listed 51 generic guidelines for conceiving, organizing, conducting and assessing ecological restoration projects that would be applicable to any habitat and land use type like farmland. The guidelines are listed under the following stages: (i) conceptual planning, i.e. identify the project site location and its boundaries (guideline no. 1) and identify restoration goals, if any, pertaining to social and cultural values (no. 5), (ii) preliminary tasks, i.e. document existing project site conditions and describe the taxa (no. 20), prepare a list of objectives designed to achieve restoration goals (no. 27), (iii) installation planning, i.e. describe the interventions that will be implemented to attain each objective (no. 34), (iv) installation tasks, i.e. implement restoration objectives (no. 42), (v) post-installation tasks, e.g. protect the project site against vandals and herbivory (no. 43), and (vi) evaluation, i.e. assess monitoring data to determine if performance standards are being met (no. 48). These guidelines and other generic principles have been supported by conceptual frameworks and practical methods for developing ecological restoration programmes and evaluating ecological risk, which are generally not specific to farmland but nevertheless provide a useful starting point for this particular land use type. It is generally accepted that practical guidelines for restoration must be based on sound theoretical and empirical foundations, yet be easily integrated into agricultural and horticultural practice. Yet in this context, made two salient remarks: firstly, practitioners now desire a stronger ecological foundation for developing and implementing restoration projects; and secondly, it has become apparent that the assumptions underlying many restoration projects are based on outdated concepts of how ecological systems function, such as on the stability of ecosystems and the ability of them to return to equilibrium states following disturbance. Indeed, it appears that some of the earlier methods applied in farmland ecological restoration, which were promoted through agri-environmental schemes, were in fact based on scant ecological empirical data, but now this situation is changing.

Before specific restoration options can be applied to a degraded area, it is necessary to consider the identity of the causal agent, the extent of present damage (i.e. condition) and the rate of change (i.e. trend). The causal agents are either biotic or abiotic factors. The appropriate response to biotic factors (e.g. grazing-induced changes in vegetation composition) should be to eliminate or reduce the degrading factor (e.g. grazing animals) and adjust the biotic composition (e.g. replant or allow to regenerate the desired species). Alternatively, the appropriate response to abiotic factors (e.g. through soil erosion or contamination) should be to initially remove the degrading factor and then repair the physical and chemical environment. In the latter case, there is little point in manipulating biotic composition if the underlying abiotic problems that impair the expected ecosystem functioning are still present.

Notwithstanding the above caveat in relation to the causal agent, ecological restoration of farmland has generally involved attempts to moderate the effects of fragmentation and habitat loss and decrease the intensity of management. Some of these methods have been supported by good empirical evidence. However, the 'ecological' component of ecological restoration was frequently lacking from some earlier programmes such as set-aside, whose foundations were instead based on untested assumptions and anecdotal observations or which restoration filled an objective subsidiary to a greater goal such as to reduce excess agricultural production. Expert opinion can be used to prioritize actions in ecological restoration. Predictive modelling approaches, including 'alternative futures' analysis, can also be used to test a variety of scenarios and are particularly useful when management planning must balance conflicting land uses. Specifically, concluded that daily ration models were the most useful type of model for simulating the response of farmland birds to changing food supply; the two broad classes of models they considered were phenomenological models (aggregative and population approaches) and behavioural depletion models (daily ration and functional response approaches).

Table: Details of generic ecological restoration methods.

S. No.	Method
1	Reduce pesticide and fertilizer use or substitute for less disruptive products in both the main crop (e.g. organics) and adjacent non-target areas like headlands.
2	Increase the size of habitat patches and connectivity between them by creating corridors or contrary to expectations, establishing small patches of vegetation to facilitate dispersal.
3	Increase the availability of non-cultivated land adjacent to fields which provide natural nesting and over-wintering sites (physical shelter and shelter from predation) and primary or alternative food sources, either by deliberately sown non-crop plants (beetle banks, cover strips, floral mixtures and hedgerows), natural regeneration (set-aside land without pesticides or conservation headlands with selective pesticides, uncropped field margins and grassy margins) or repair of existing vegetation.
4	Increasing the number of fields by reducing the size of each.
5	Establish artificial nests or shelter and feeding stations.
6	Substitute or diversify the species of arable crops grown at any one time and over the year (arable reversion to pasture, crop rotation, retain over-wintered stubble in a spring fallow, spring not autumn sown cereals such as winter wheat, undersow crops with grasses).
7	Stagger the timing and location of adverse husbandry practices such as soil cultivation, planting and harvesting (strip harvesting).
8	Translocate animals despite the high cost over large areas.
9	Remove unwanted animals and plants, including fencing to exclude stock from sensitive riparian areas.

A few recent examples of empirical restoration research designed to support the above farmland ecological restoration methods are provided below. Recent work in New Zealand is using untreated discs of pine wood to accelerate ecological succession; these discs provide many of the ecological functions of natural fallen logs and can harbour late-succession invertebrate communities which usually would not be present in highly modified farming landscapes. A similar example from New Zealand, which again involves designing end of succession habitats, is the use of 'Weta Hotels'. Wetas (Orthoptera: Anostostomatidae and Rhaphidophoridae) are large, iconic native insects that are usually associated with undisturbed forest landscapes. Weta hotels, artificial shelters which mimic coarse woody debris with cavities, harbour this specialist fauna on farmland on which this insect disappeared in association with forest clearance. Populations of pollinating bees and predatory wasps in agroecosystems can also be enhanced with the introduction of suitable nesting sites.

The provision of non-crop vegetation as a food source for immature and adult insect natural enemies and birds requires a good understanding of food-web theory. The 'shotgun' approach of planting a species-rich wildflower/wildlife seed mixture or by allowing natural regeneration of species from the seed bank may result in inadvertent proliferation (and possible dominance) of a single or few species in the target area. Ill-conceived selection of flowering plants could unintentionally enhance the incidence of herbivorous pests, higher-order predators/hyperparasitoids or plant diseases. This could lead to disillusionment among agricultural and horticultural practitioners with habitat restoration methods. In order for these methods to be effective, their implementation must be guided by empirical and theoretical research, such as the initial screening of individual tussock grass or flower species alone and then in mixtures.

The monitoring of restoration projects to judge their success generally involves consideration of three major ecosystem attributes: (i) diversity (species richness, evenness and abundance), (ii) vegetation structure (cover, density or biomass), and (iii) ecological processes (e.g. nutrient cycling, rates of herbivory and predation. Importantly, these three attributes are not only evaluated in the restored site, but also in unrestored ('impacted') and intact ('unimpacted') reference sites to evaluate the level of restoration success against the nominal objectives. Further, the restoration 'treatments' should be replicated across different sites. The SERI has produced a 'primer' that provides a list of nine ecosystem attributes as a guideline for measuring restoration success. They suggested that a restored ecosystem should have the following attributes: (i) similar diversity and community structure in comparison with reference sites, (ii) presence of indigenous species, (iii) presence of functional groups necessary for long-term stability, (iv) capacity of the physical environment to sustain reproducing populations, (v) normal functioning, (vi) integration with the landscape, (vii) elimination or reduction of potential threats, (viii) resilience to natural disturbances, and (ix) self-sustaining. Modelling approaches can be used to evaluate restoration success. Thus, ecological restoration of farmland involves adapting guidelines, applying methods and assessing the success against the objectives, and if possible, reference sites.

Agri-environmental Schemes

Public concern over the environmental impacts of agriculture has been a significant driver for the introduction of various types of agri-environmental scheme. A defining characteristic is that they provide compensation to farmers for any loss of income that results from their implementation of prescribed measures designed to benefit the environment: 'payment for habitat and wildlife gain'.

The first agri-environmental scheme was introduced in the United Kingdom in 1986 and a series of schemes has since been introduced: the Environmentally Sensitive Area Scheme, the Countryside Stewardship Scheme, the Arable Stewardship Scheme and most recently the England Rural Development Program. Support from the European Union (EU) Common Agricultural Policy (CAP) has led to agri-environmental schemes becoming popular within and outside the EU with schemes in place in at least 26 of the 44 European countries including Belgium, Finland, Scotland and Sweden. The Conservation Reserve Program occupies a similar place in the USA.

Ecological Background

Quantitative estimates available of the economic value of biodiversity are supported by ecological theory. Though the science of valuing and mechanistically understanding ecosystem services is in its infancy, empirical studies tend to support the notion that increasing biodiversity favours ecosystem functions of value such as biological control of crop pests, pollination and nutrient cycling. However, the ecosystem function–biodiversity debate is highly topical among ecologists and the circumstances under which ecosystem function actually is enhanced can be complex, and sometimes counter-intuitive. Typically, the response function to biodiversity increase takes the form of a linear or saturation curve. Though the addition of a single species to an already complex ecosystem is likely to have only a small effect (the saturation curve), this is less likely to apply in simple agroecosystems than in more complex natural systems. The accelerating form of response is not well supported by empirical data and this has important practical implications. If the response were of this type then harnessing the ecosystem benefits of biodiversity would be far less tractable owing to the high numbers of species and microhabitats, as well as land area and genetic diversity that would need to be conserved or reinstated. Evidence that relatively large increases in ecosystem function may be achieved with low or modest levels of biodiversity enhancement supports the potential use of agri-environmental schemes for the provision of ecosystem services as well as for altruistic reasons (e.g. conservation of endangered but not directly useful species) and political reasons (e.g. reducing agricultural overproduction). Indeed, using agri-environmental schemes to promote numerous ecosystem services from restored farmland is encapsulated in the blueprint proposal for a 'multifunctional' countryside.

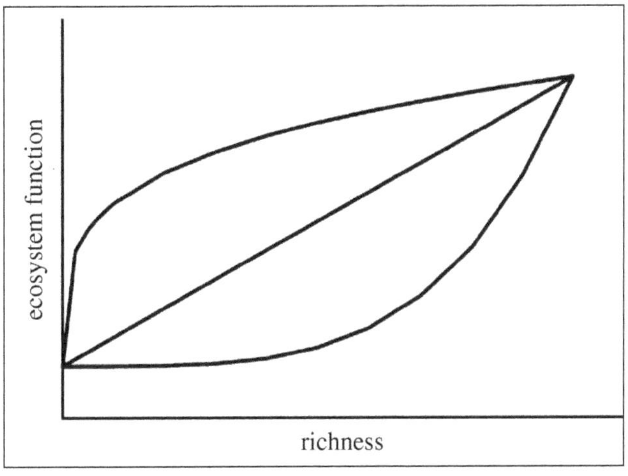

Theoretical relationships between species richness and ecosystem function. The top line shows saturation in which the majority of species contribute little, middle line shows a linear relationship in which each additional species adds a consistent unit of function and bottom line shows an accelerating function whereby interactions between species enhance their efficiencies. The latter is least well supported by empirical data.

An increasingly important aspect of biodiversity in agricultural systems is that many species are exotic, especially in Australasia. Quite rightly, a great deal of attention has been given to the negative effects of such introduced species on endemics but relatively little attention has been given to the potentially positive effects on ecosystem services. Ecological fitting occurs when a species expands its geographical range and enters into novel ecological associations; for example, hummingbird use of exotic plants. The most remarkable example is the development of an apparently fully functional cloud rainforest of exotic plant species on the mid-Atlantic island of Ascension, a location reported to be destitute of trees when Darwin visited in 1836. Ecological fitting is apparent also in agricultural systems in cases where exotic plants are recognized and used by native natural enemies as in the case of the Australian endemic egg parasitoid Trichogramma carverae using the Mediterranean plant Lobularia maritima (L.) Desv. (sweet alyssum) as a nectar source with consequent increases in their parasitism of vineyard pests. Other similar examples are known from the conservation biological control literature suggesting that closer examination may reveal cases of services other than biological control involving ecological fitting. This concept further supports the value of biodiversity being enhanced within agri-environmental schemes despite the apparently unavoidable presence of exotics.

Costs and Benefits

In the EU where agri-environmental schemes are most widely used, only a small fraction of total government expenditure on the CAP was historically allocated to the support of the schemes—4% in 1999. Of course this is a small part of a large overall budget; accordingly, total expenditure on EU agri-environmental schemes since 1994 has been estimated at €24.3 billion. Recently, the EU's new Rural Development Regulation made possible the allocation of up to 20% of CAP funding to agri-environmental and rural development schemes. Such levels of expenditure have made farmers in some countries like Austria heavily dependent on this funding. The conditions placed upon farmers in order to qualify for payments may be rigorous. In Switzerland, it is a requirement that at least 7% of every farm is allocated to ecological compensation areas.

Though there have been many attempts to measure the effectiveness of agri-environmental schemes, concluded that the research design was inadequate in the majority and almost one-third did not include a statistical analysis of data. The most common fault was that most studies compared the areas subject to agri-environmental schemes with control areas, despite the fact that areas chosen for schemes often were superior from the outset, that being part of the basis for their protection and enhancement. Some studies avoided this problem by collecting baseline data and comparing trends or changes or more careful selection of comparison sites.

In a meta-analysis of agri-environmental schemes, evaluations that covered birds, arthropods and plants, just over half of all taxa exhibited increases in richness or abundance and only 6% declined in comparison with control areas. Though these overall statistics appear encouraging, there is a degree of dissatisfaction that agri-environmental schemes are failing to achieve their conservation aims. In Ireland, there appeared to be a need for better farmer education and training and in the

Netherlands the high overall intensity of land use was thought to constrain improvements. Other studies, however, show encouraging signs of reversal of long established trends in wildlife decline in the UK or increases in Swiss biodiversity.

Improving Agri-environmental Schemes

Given the criticism directed to the mixed success of agri-environmental schemes and of the methods used to evaluate these impacts, there is a need to improve not only our understanding of the indicators to be used in assessment of schemes but also to involve ecologists to a greater extent in the development and measurement of schemes. An example of how theory may support the design of schemes is provided by the 'habitat creation model' to identify those vegetation communities that can be established on ex-arable land. A further ecological consideration is the scale over which processes occur. Increasingly, there is empirical evidence that landscape-level factors affect ecosystem service providers such as pollinators and natural enemies. Such effects may be especially important at large temporal scales, i.e. effects over more than one cropping phase. It is therefore critical that agri-environmental schemes address the importance of landscape-scale effects, including the relationship between local and regional diversity. In practical terms, this could take the form of better orchestration above the level of individual farms to ensure that the implementation of agri-environmental schemes within a landscape is optimal in relation to each other and the existence of other significant features like woodland. Doing this will, however, demand better knowledge of how ecosystem service providers respond to habitat fragmentation (and measures for its amelioration such as wildlife corridors) and habitat loss (and measures for its amelioration such as taking parcels of land out of production for afforestation and other habitat). This may not be as simple as it may appear first. In the case of natural enemies of pests, for example, research has shown that beneficial species may respond at spatial scales different from those applying to the pests. Though this finding signals tantalizing scope to 'engineer' landscapes to favour beneficials over pests, important differences may occur even within beneficial groups. For example, parasitoids responded to fragmentation of forest elements at differing scales, an effect apparently linked to the body size of the insects.

An emerging trend for agri-environmental schemes is 'multifunctionality' rather than narrowly defined environmental objectives. Accordingly, the multiple objectives must be reflected in a multidisciplinary approach to evaluation as advocated by Carey et al. An important consideration for future schemes is to shift financial support from agricultural production to 'stewardship' activities that will achieve desired environmental outcomes, a 'post-productionist' paradigm in the terms of Wilson. There is little research available on how production and stewardship imperatives may be balanced in countries outside the EU. In Australia, for example, there has been a history of very little State or Commonwealth production support. Accordingly, the rather limited agri-environmental scheme-type measures that have been used, the 'Landcare' movement for example, have been largely independent of production incentives.

Farmer attitudes towards agri-environmental schemes clearly play a major role in uptake, the extent to which guidelines are followed and the degree to which outcomes are achieved. A wide spectrum of degree of commitment and sympathy was evident in one survey of British farmers and these fell into two approximately equal sized categories. 'Passive adopters' were driven primarily by financial rewards and appeared adept at maximizing payments while minimizing effort or impact on normal farming activities by, for example, 'resting' poor quality arable land. One interviewee commented, 'Basically it is attractive because you are getting paid for what you are already doing'. Fortunately, 49% of subjects were 'active adopters' with an appreciation of the objectives of agri-environmental schemes and willing to make changes to pursue environmental objectives. Attracting farmers with such attitudes into schemes was considered an important indicator of success but the fact that participants in schemes are self-selecting reinforces the need for careful design of schemes in terms of eligibility requirements and implementation and compliance regulations. A related problem associated with agri-environmental schemes is the confusion caused by multiple schemes in a given jurisdiction and the sheer complexity of the regulations associated with each. This factor led to propose a 'one-stop shop' for farmers to access information on the schemes for which they may be eligible.

Issues such as those dealt with above illustrate the need for social, as well as ecological, science inputs for the design, implementation and evaluation of agri-environmental schemes. Success is influenced by human as well as technical factors.

AGROECOSYSTEM ANALYSIS

Agroecosystem analysis is a thorough analysis of an agricultural environment which considers aspects from ecology, sociology, economics, and politics with equal weight. There are many aspects to consider; however, it is literally impossible to account for all of them. This is one of the issues when trying to conduct an analysis of an agricultural environment. In the past, an agroecosystem analysis approach might be used to determine the sustainability of an agricultural system. It has become apparent, however, that the "sustainability" of the system depends heavily on the definition of sustainability chosen by the observer. Therefore, agroecosystem analysis is used to bring the richness of the true complexity of agricultural systems to an analysis to identify reconfigurations of the system (or *holon*) that will best suit individual situations.

Agroecosystem analysis is a tool of the multidisciplinary subject known as Agroecology. Agroecology and agroecosystem analysis are not the same as sustainable agriculture, though the use of agroecosystem analysis may help a farming system ensure its viability. Agroecosystem analysis is not a new practice, agriculturalists and farmers have been doing it since societies switched from hunting and gathering (hunter-gatherer) for food to settling in one area. Every time a person involved in agriculture evaluates their situation to identify methods to make the system function in a way that better suits their interests, they are performing an agroecosystem analysis.

Agroecosystem Analysis and Sustainable Agriculture Differ

It is difficult to discuss these differences without the aid of an example. Consider the case of a conventional apple farmer. This farmer may choose to change his farm to conform to the standards of USDA approved organic agriculture because he felt motivated by social or moral norms or the potential of increased profits or a host of other reasons. This farmer evaluated his situation and reconfigured it to try to improve it. Some might look at this situation and conclude that the apple farmer chose organic apple production because it is more sustainable for the environment. But, what if a few years later the farmer finds that he is struggling to make a profit and decides to go back to conventional agriculture? The farmer performed another agroecosystem analysis and arrived at a reconfiguration that some might see as unsustainable. This example illustrates how agroecosystem analysis is not required to lead a more environmentally sustainable form of agriculture. Agroecosystem analysis might produce a reconfiguration that is more economically sustainable or socially sustainable or politically sustainable for a farmer (or other actor). By definition, however, agroecosystem analysis is not required to produce an environmentally sustainable configuration for an agricultural system.

Approach to Analysis

William L. Bland, from the University of Wisconsin–Madison, developed the idea of a farm as a Holon. This term, holon, was originally introduced by Arthur Koestler in 1966, in which he referred to a holon as an entity in which it is a part by itself, a holon, while contributing to a larger entity, which is also a holon. Bland develops this for an agricultural environment or farm as, "The farm holon is both the whole in which smaller holons exists, and a part of larger entities, themselves holons." This idea was expanded upon by Bland and Michael M. Bell University of Wisconsin–Madison in 2007 "A holon approach to agroecology," because it is difficult to account for boundary and change when using a systems thinking approach. One major difference between Koestler's holon and the holon idea developed for agroecosystem analysis is that the latter can only be defined as a holon if it has intentionality.

The farm itself is a holon and within the farm holon, other holons exist. For example, a farm animal, the farm family, and a farmworker can all be considered holons within the farm. Additionally, the farm is considered a holon which is inpart connected to other holons such as the county in which the farm resides, the bank from which the farmer borrowed money, or the grain elevator where the farmer can sell goods. Things like the tractor or the barn are not holons because they lack intentionality.

When conducting an agroecosystem analysis, the analyst should approach the farm as the farm itself and the "ecology of contexts" in which the farm and the farmer function. A "context" is anything that might influence functioning of the farm and cause it to change. According to Bland and Bell, examples of contexts include, "family, farm business, genetic heart disease, and spiritual beliefs." These examples illustrate the breadth of contexts that could influence why farmers do what they do. Bland concluded his model of a farm as a *holon* by stating, "A farm is not sustainable (disintegrates) when it cannot find an overall configuration that is simultaneously viable in all contexts."

Questions to Consider

There is no right or wrong way to evaluate an agroecosystem. It is important to identify all actors in a holon before beginning the analysis. When an analyst accepts the task of analyzing the agroecosystem, first and foremost, it must be approached as to incorporate all elements involved and should derive questions that should be answered. Questions such as:

- What defining factors (holons and contexts) determine the present configuration of the agroecosystem?
- How does one quantify the sustainability of the farm holon (economic, social, political, ecological and other)?
- How does the farmer or farm family perceive an agroecosystem?
- What is the farmer doing now, and how do those practices or actions affect the viability of the agroecosystem?

- Can the farmer maintain his livelihood continuing with current practices?
- What does the farmer value and where do those values come from?
- Will the farmer consider alternative farm configurations?

These are the types of questions an analyst could consider. There are no preset questions to ask, and usually more questions are derived than answered. However, the most important task an analysts can do, is to start the analysis with an open mind and under no presumptions about what is and is not sustainable for the farm holon.

Analysis Types

J. Visser of Dordt College uses a diagram, "Wealth Creation Wheel" to emphasize and account for the parameters of developing a thorough analysis. His diagram is more emphasized on economics; however, it is a useful tool to reference when starting to analyze an agroecoystem. His interest is to create a functioning wheel which will *roll* when all parameters are met equally. If one parameter is not functioning in context with the other parameters, then the wheel will be out of balance and ineffective, thus unsustainable. When referring to an agroecosystem, if one parameter is out of balance, this could lead to an unproductive cropping season and loss of income and livelihood.

MICROBIAL IMPACT ON AGROECOSYSTEM

Nutrient Cycling

A nutrient cycle (or ecological recycling) is the movement and exchange of organic and inorganic matter back into the production of matter. Energy flow is a unidirectional and noncyclic pathway, whereas the movement of mineral nutrients is cyclic. Mineral cycles include the carbon cycle, sulfur cycle, nitrogen cycle, water cycle, phosphorus cycle, oxygen cycle, among others that continually recycle along with other mineral nutrients into productive ecological nutrition.

Fallen logs are critical components of the nutrient cycle in terrestrial forests. Nurse logs form habitats for other creatures that decompose the materials and recycle the nutrients back into production.

The nutrient cycle is nature's recycling system. All forms of recycling have feedback loops that use energy in the process of putting material resources back into use. Recycling in ecology is regulated to a large extent during the process of decomposition. Ecosystems employ biodiversity in the food webs that recycle natural materials, such as mineral nutrients, which includes water. Recycling in natural systems is one of the many ecosystem services that sustain and contribute to the well-being of human societies.

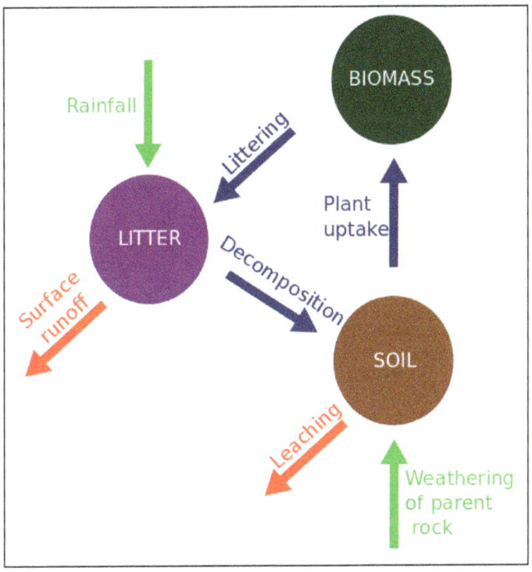

A nutrient cycle of a typical terrestrial ecosystem.

There is much overlap between the terms for the biogeochemical cycle and nutrient cycle. Most textbooks integrate the two and seem to treat them as synonymous terms. However, the terms often appear independently. Nutrient cycle is more often used in direct reference to the idea of an intra-system cycle, where an ecosystem functions as a unit. From a practical point, it does not make sense to assess a terrestrial ecosystem by considering the full column of air above it as well as the great depths of Earth below it. While an ecosystem often has no clear boundary, as a working model it is practical to consider the functional community where the bulk of matter and energy transfer occurs. Nutrient cycling occurs in ecosystems that participate in the "larger biogeochemical cycles of the earth through a system of inputs and outputs."

Complete and Closed Loop

- All systems recycle: The biosphere is a network of continually recycling materials and information in alternating cycles of convergence and divergence. As materials converge or become more concentrated they gain in quality, increasing their potentials to drive useful work in proportion to their concentrations relative to the environment. As their potentials are used, materials diverge, or become more dispersed in the landscape, only to be concentrated again at another time and place.
- Ecosystems are capable of complete recycling: Complete recycling means that 100% of the waste material can be reconstituted indefinitely. This idea was captured by Howard T. Odum when he penned that "it is thoroughly demonstrated by ecological systems and geological systems that all the chemical elements and many organic substances can be

accumulated by living systems from background crustal or oceanic concentrations without limit as to concentration so long as there is available solar or another source of potential energy". In 1979 Nicholas Georgescu-Roegen proposed the fourth law of entropy stating that complete recycling is impossible. Despite Georgescu-Roegen's extensive intellectual contributions to the science of ecological economics, the fourth law has been rejected in line with observations of ecological recycling. However, some scientists state that complete recycling is impossible for technological waste.

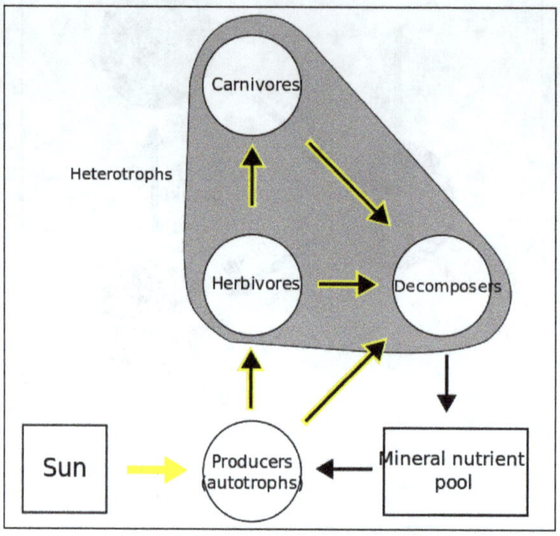

A simplified food web illustrating a three-trophic food chain (producers-herbivores-carnivores) linked to decomposers. The movement of mineral nutrients through the food chain, into the mineral nutrient pool, and back into the trophic system illustrates ecological recycling. The movement of energy, in contrast, is unidirectional and noncyclic.

Ecosystems execute closed loop recycling where demand for the nutrients that adds to the growth of biomass exceeds supply within that system. There are regional and spatial differences in the rates of growth and exchange of materials, where some ecosystems may be in nutrient debt (sinks) where others will have extra supply (sources). These differences relate to climate, topography, and geological history leaving behind different sources of parent material. In terms of a food web, a cycle or loop is defined as "a directed sequence of one or more links starting from, and ending at, the same species." An example of this is the microbial food web in the ocean, where "bacteria are exploited, and controlled, by protozoa, including heterotrophic microflagellates which are in turn exploited by ciliates. This grazing activity is accompanied by excretion of substances which are in turn used by the bacteria so that the system more or less operates in a closed circuit."

Ecological Recycling

A large fraction of the elements composing living matter reside at any instant of time in the world's biota. Because the earthly pool of these elements is limited and the rates of exchange among the various components of the biota are extremely fast with respect to geological time, it is quite evident that much of the same material is being incorporated again and again into different biological forms. This observation gives rise to the notion that, on the average, matter (and some amounts of energy) are involved in cycles.

An example of ecological recycling occurs in the enzymatic digestion of cellulose. "Cellulose, one of the most abundant organic compounds on Earth, is the major polysaccharide in plants where it is part of the cell walls. Cellulose-degrading enzymes participate in the natural, *ecological recycling* of plant material." Different ecosystems can vary in their recycling rates of litter, which creates a complex feedback on factors such as the competitive dominance of certain plant species. Different rates and patterns of ecological recycling leaves a legacy of environmental effects with implications for the future evolution of ecosystems.

Ecological recycling is common in organic farming, where nutrient management is fundamentally different compared to agri-business styles of soil management. Organic farms that employ ecosystem recycling to a greater extent support more species (increased levels of biodiversity) and have a different food web structure. Organic agricultural ecosystems rely on the services of biodiversity for the recycling of nutrients through soils instead of relying on the supplementation of synthetic fertilizers. The model for ecological recycling agriculture adheres to the following principals:

- Protection of biodiversity,
- Use of renewable energy,
- Recycling of plant nutrients.

Where produce from an organic farm leaves the farm gate for the market the system becomes an open cycle and nutrients may need to be replaced through alternative methods.

Ecosystem Engineers

An illustration of an earthworm casting.

The persistent legacy of environmental feedback that is left behind by or as an extension of the ecological actions of organisms is known as niche construction or ecosystem engineering. Many species leave an effect even after their death, such as coral skeletons or the extensive habitat modifications to a wetland by a beaver, whose components are recycled and re-used by descendants and other species living under a different selective regime through the feedback and agency of these legacy effects. Ecosystem engineers can influence nutrient cycling efficiency rates through their actions.

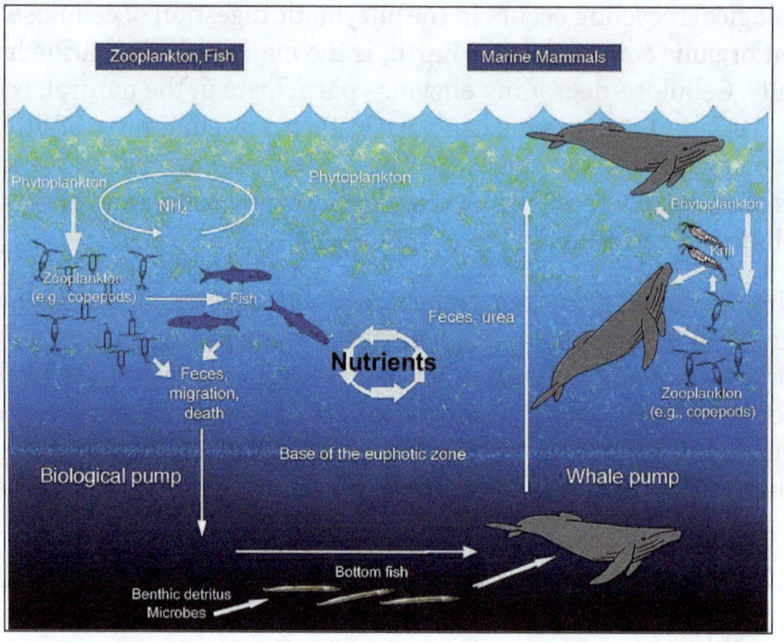

From the largest to the smallest of creatures, nutrients are recycled by their movement, by their wastes, and by their metabolic activities. This illustration shows an example of the whale pump that cycles nutrients through the layers of the oceanic water column. Whales can migrate to great depths to feed on bottom fish (such as sand lance Ammodytes spp.) and surface to feed on krill and plankton at shallower levels. The whale pump enhances growth and productivity in other parts of the ecosystem.

Earthworms, for example, passively and mechanically alter the nature of soil environments. Bodies of dead worms passively contribute mineral nutrients to the soil. The worms also mechanically modify the physical structure of the soil as they crawl about (bioturbation), digest on the molds of organic matter they pull from the soil litter. These activities transport nutrients into the mineral layers of soil. Worms discard wastes that create worm castings containing undigested materials where bacteria and other decomposers gain access to the nutrients. The earthworm is employed in this process and the production of the ecosystem depends on their capability to create feedback loops in the recycling process.

Shellfish are also ecosystem engineers because they: 1) Filter suspended particles from the water column; 2) Remove excess nutrients from coastal bays through denitrification; 3) Serve as natural coastal buffers, absorbing wave energy and reducing erosion from boat wakes, sea level rise and storms; 4) Provide nursery habitat for fish that are valuable to coastal economies.

Fungi contribute to nutrient cycling and nutritionally rearrange patches of ecosystem creating niches for other organisms. In that way fungi in growing dead wood allow xylophages to grow and develop and xylophages, in turn, affect dead wood, contributing to wood decomposition and nutrient cycling in the forest floor.

Nitrogen Fixation

Nitrogen fixation is a process by which molecular nitrogen in the air is converted into ammonia (NH_3) or related nitrogenous compounds in soil. Atmospheric nitrogen is molecular dinitrogen, a relatively nonreactive molecule that is metabolically useless to all but a few microorganisms. Biological nitrogen fixation converts N_2 into ammonia, which is metabolized by most organisms.

Nitrogen fixation is essential to life because fixed inorganic nitrogen compounds are required for the biosynthesis of all nitrogen-containing organic compounds, such as amino acids and proteins, nucleoside triphosphates and nucleic acids. As part of the nitrogen cycle, it is essential for agriculture and the manufacture of fertilizer. It is also, indirectly, relevant to the manufacture of all nitrogen chemical compounds, which includes some explosives, pharmaceuticals and dyes.

Nitrogen fixation is carried out naturally in soil by microorganisms termed diazotrophs that include bacteria such as *Azotobacter* and archaea. Some nitrogen-fixing bacteria have symbiotic relationships with plant groups, especially legumes. Looser non-symbiotic relationships between diazotrophs and plants are often referred to as associative, as seen in nitrogen fixation on rice roots. Nitrogen fixation occurs between some termites and fungi. It occurs naturally in the air by means of NOx production by lightning.

All biological nitrogen fixation is effected by enzymes called nitrogenases. These enzymes contain iron, often with a second metal, usually molybdenum but sometimes vanadium.

Fixation

Biological

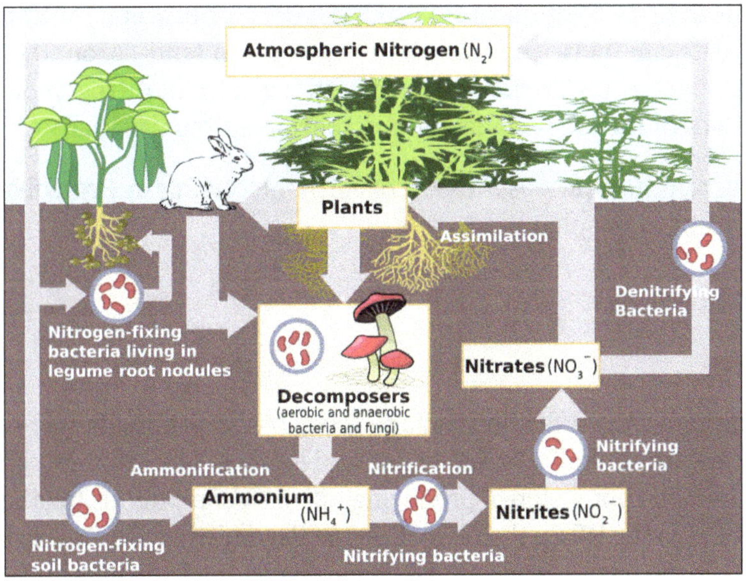

Schematic representation of the nitrogen cycle. Abiotic nitrogen fixation has been omitted.

Biological nitrogen fixation was discovered by German agronomist Hermann Hellriegel and Dutch microbiologist Martinus Beijerinck. Biological nitrogen fixation (BNF) occurs when atmospheric nitrogen is converted to ammonia by a nitrogenase enzyme. The overall reaction for BNF is:

$$N_2 + 16ATP + 8e^- + 8H^+ \rightarrow 2NH_3 + H_2 + 16ADP + 16P_i$$

The process is coupled to the hydrolysis of 16 equivalents of ATP and is accompanied by the co-formation of one equivalent of H_2. The conversion of N_2 into ammonia occurs at a metal cluster called

FeMoco, an abbreviation for the iron-molybdenum cofactor. The mechanism proceeds via a series of protonation and reduction steps wherein the FeMoco active site hydrogenates the N_2 substrate. In free-living diazotrophs, nitrogenase-generated ammonia is assimilated into glutamate through the glutamine synthetase/glutamate synthase pathway. The microbial nif genes required for nitrogen fixation are widely distributed in diverse environments.

Nitrogenases are rapidly degraded by oxygen. For this reason, many bacteria cease production of the enzyme in the presence of oxygen. Many nitrogen-fixing organisms exist only in anaerobic conditions, respiring to draw down oxygen levels, or binding the oxygen with a protein such as leghemoglobin.

Microorganisms

Diazotrophs are widespread within domain Bacteria including cyanobacteria (e.g. the highly significant Trichodesmium and Cyanothece), as well as green sulfur bacteria, Azotobacteraceae, rhizobia and Frankia. Several obligately anaerobic bacteria fix nitrogen including many (but not all) Clostridium spp. Some archaea also fix nitrogen, including several methanogenic taxa, which are significant contributors to nitrogen fixation in oxygen-deficient soils.

Cyanobacteria inhabit nearly all illuminated environments on Earth and play key roles in the carbon and nitrogen cycle of the biosphere. In general, cyanobacteria can use various inorganic and organic sources of combined nitrogen, such as nitrate, nitrite, ammonium, urea, or some amino acids. Several cyanobacteria strains are also capable of diazotrophic growth, an ability that may have been present in their last common ancestor in the Archean eon. Nitrogen fixation by cyanobacteria in coral reefs can fix twice as much nitrogen as on land—around 660 kg/ha/year. The colonial marine cyanobacterium Trichodesmium is thought to fix nitrogen on such a scale that it accounts for almost half of the nitrogen fixation in marine systems globally.

Marine surface licing and non-photosynthetic bacteria belonging in Proteobacteria and Planctomycetes fixate significant atmospheric nitrogen.

Root Nodule Symbioses

Legume Family

Plants that contribute to nitrogen fixation include those of the legume family—Fabaceae, with taxa such as kudzu, clover, soybean, alfalfa, lupin, peanut and rooibos. They contain symbiotic rhizobia bacteria within nodules in their root systems, producing nitrogen compounds that help the plant to grow and compete with other plants. When the plant dies, the fixed nitrogen is released, making it available to other plants; this helps to fertilize the soil. The great majority of legumes have this association, but a few genera (e.g., Styphnolobium) do not. In many traditional and organic farming practices, fields are rotated through various types of crops, which usually include one consisting mainly or entirely of clover or buckwheat (non-legume family Polygonaceae), often referred to as "green manure".

Fixation efficiency in soil is dependent on many factors, including the legume and air and soil conditions. For example, nitrogen fixation by red clover can range from 50-200 lb/acre.

Inga alley farming relies on the leguminous genus Inga, a small tropical, tough-leaved, nitrogen-fixing tree.

Non-leguminous

A sectioned alder tree root nodule.

Other nitrogen fixing families include:

- Parasponia, a tropical genus in the family Cannabaceae, which are able to interact with rhizobia and form nitrogen-fixing nodules.
- Actinorhizal plants such as alder and bayberry can form nitrogen-fixing nodules, thanks to a symbiotic association with Frankia bacteria. These plants belong to 25 genera distributed across eight families.

The ability to fix nitrogen is present in other families that belong to the orders Cucurbitales, Fagales and Rosales, which together with the Fabales form a clade of eurosids. The ability to fix nitrogen is not universally present in these families. For example, of 122 Rosaceae genera, only four fix nitrogen. Fabales were the first lineage to branch off this nitrogen-fixing clade; thus, the ability to fix nitrogen may be plesiomorphic and subsequently lost in most descendants of the original nitrogen-fixing plant; however, it may be that the basic genetic and physiological requirements were present in an incipient state in the most recent common ancestors of all these plants, but only evolved to full function in some of them.

Family: Genera				
Betulaceae: Alnus (alders)	Coriariaceae: Coriaria	Myricaceae	Rhamnaceae	Rosaceae
Cannabaceae: Trema	Datiscaceae: Datisca	Comptonia (sweetfern)	Ceanothus	Cercocarpus (mountain mahoganies)
Casuarinaceae	Elaeagnaceae	Morella	Colletia	Chamaebatia (mountain miseries)
Allocasuarina	Elaeagnus (silverberries)	Myrica (bayberries)	Discaria	Dryas
Casuarina	Hippophae (sea-buckthorns)		Kentrothamnus	Purshia/Cowania (bitterbrushes/cliffroses)
Ceuthostoma	Shepherdia (buffaloberries)		Retanilla	
Gymnostoma			Talguenea	
			Trevoa	

Several nitrogen-fixing symbiotic associations involve cyanobacteria (such as *Nostoc*):

- Some lichens such as Lobaria and Peltigera,
- Mosquito fern (Azolla species),
- Cycads,
- Gunnera,
- Blasia (liverwort),
- Hornworts.

Endosymbiosis in Diatoms

Rhopalodia gibba, a diatom alga, is a eukaryote with cyanobacterial N_2-fixing endosymbiont organelles. The spheroid bodies reside in the cytoplasm of the diatoms and are inseparable from their hosts.

SOIL ORGANIC MATTER

Soil organic matter (SOM) is the organic matter component of soil, consisting of plant and animal detritus at various stages of decomposition, cells and tissues of soil microbes, and substances that soil microbes synthesize. SOM provides numerous benefits to the physical and chemical properties of soil and its capacity to provide regulatory ecosystem services. SOM is especially critical for soil functions and quality.

The benefits of SOM result from a number of complex, interactive, edaphic factors; a non-exhaustive list of these benefits to soil function includes improvement of soil structure, aggregation, water retention, soil biodiversity, absorption and retention of pollutants, buffering capacity, and the cycling and storage of plant nutrients. SOM increases soil fertility by providing cation exchange sites and being a reserve of plant nutrients, especially nitrogen (N), phosphorus (P), and sulfur (S), along with micronutrients, which the mineralization of SOM slowly releases. As such, the amount of SOM and soil fertility are significantly correlated.

SOM also acts as a major sink and source of soil carbon (C). Although the C content of SOM varies considerably, SOM is ordinarily estimated to contain 58% C, and "soil organic carbon" (SOC) is often used as a synonym for SOM, with measured SOC content often serving as a proxy for SOM. Soil represents one of the largest C sinks on Earth and is significant in the global carbon cycle. Therefore, SOM/SOC dynamics and the capacity of soils to provide the ecosystem service of carbon sequestration through SOM management have received considerable attention recently.

The concentration of SOM in soils generally ranges from 1% to 6% of the total mass of topsoil for most upland soils. Soils whose upper horizons consist of less than 1% of organic matter are mostly limited to deserts, while the SOM content of soils in low lying, wet areas can be as great as 90%. Soils containing 12% to 18% SOC are generally classified as organic soils.

It can be divided into 3 genera: the living biomass of microbes, fresh and partially decomposed detritus, and humus. Surface plant litter, i.e. fresh vegetal detritus, is generally excluded from SOM.

Sources of SOM

The primary source of SOM is vegetal detritus. In forests and prairies, for example, different organisms decompose the fresh detritus into simpler compounds. This involves several stages, the first being mostly mechanical, and becoming more chemical as decomposition progresses. The microbial decomposers are included in the SOM, and form a food web of organisms that prey upon each other and subsequently become prey.

There are also other herbivores that consume fresh vegetal matter, the residue of which then passes to the soil. The products of the metabolisms of these organisms are the secondary sources of SOM, which also includes their corpses. Some animals, like earthworms, ants, and centipedes contribute to both vertical and horizontal translocation of organic matter. Additional sources of SOM include plant root exudates and charcoal.

Composition of Vegetal Detritus

The water content of most vegetal detritus is in the range of 60% to 90%. The dry matter consists of complex organic matter that is composed primarily of carbon, oxygen, and hydrogen. Although these three elements make up about 92% of the dry weight of the organic matter in soil, other elements are very important for the nutrition of plants, including nitrogen, phosphorus, potassium, sulfur, calcium, magnesium, and many micronutrients.

Organic compounds in vegetal detritus include:

- Carbohydrates that are composed of carbon, hydrogen, and oxygen, and range in complexity from rather simple sugars to the large molecules of cellulose.
- Fats that are composed of glycerids of fatty acids, like butyric, stearic, and oleic. They also include carbon, oxygen, and hydrogen.
- Lignins that are complex compounds, form the older parts of wood, and also are composed primarily of carbon, oxygen, and hydrogen. They are resistant to decomposition.
- Proteins that include nitrogen in addition to carbon, hydrogen, and oxygen; and small amounts of sulfur, iron, and phosphorus.
- Charcoal, which is elemental carbon that is derived from incomplete combustion of organic matter. It is resistant to decomposition.

Decomposition

Vegetal detritus in general is not soluble in water and therefore is inaccessible to plants. It constitutes, nevertheless, the raw matter from which plant nutrients derive. Soil microbes decompose it through enzymatic biochemical processes, obtain the necessary energy from the same matter, and produce the mineral compounds that plant roots are apt to absorb. The decomposition of organic compounds specifically into mineral, i.e., inorganic, compounds is denominated “mineralization”.

A portion of organic matter is not mineralized and instead decomposed into stable organic matter that is denominated "humus".

The decomposition of organic compounds occurs at very different rates, depending on the nature of the compound. The ranking, from fast to slow rates, is:

- Sugars, starches, and simple proteins,
- Proteins,
- Hemicelluloses,
- Cellulose,
- Lignins and fats.

The reactions that occur can be included in one of 3 genera:

- Enzymatic oxidation that produces carbon dioxide, water, and heat. It affects the majority of the matter.
- A series of specific reactions liberates and mineralizes the essential elements nitrogen, phosphorus, and sulfur.
- Compounds that are resistant to microbial action are formed by modification of the original compounds or by microbial synthesis of new ones to produce humus.

Table: The mineral products.

Element	Mineral Products
Carbon	CO_2, CO_3^{2-}, HCO_3^-, CH_4, C
Nitrogen	NH_4^+, NO_2^-, NO_3^-, N_2 (gas), N_2O (gas)
Sulfur	S, H_2S, SO_3^{2-}, SO_4^{2-}, CS_2
Phosphorus	$H_2PO_4^-$, HPO_4^{2-}
Others	H_2O, O_2, H_2, H^+, OH^-, K^+, Ca^{2+}, Mg^{2+}, etc.

Humus

As vegetal detritus decomposes, some microbially resistant compounds are formed, including modified lignins, oils, fats, and waxes. Secondly, some new compounds are synthesized, like polysaccharides and polyuronids. These compounds are the basis of humus. New reactions occur between these compounds and some proteins and other products that contain nitrogen, thus incorporating nitrogen and avoiding its mineralization. Other nutrients are also protected in this way from mineralization.

Humic Acids

Humic acids or substances are classified into 3 genera based on their solubility in acids and alkalis, and also according to their stability:

- Fulvic acid is the genus that contains the matter that has the lowest molecular weight, is soluble in acids and alkalis, and is susceptible to microbial action.

- Humic acid is the genus that contains the intermediate matter that has medial molecular weight, is soluble in alkalis and insoluble in acids, and has some resistance to microbial action.
- Humin is the genus that contains the matter that has the greatest molecular weight, is the darkest in color, is insoluble in acids and alkalis, and has the greatest resistance to microbial action.

Function in Carbon Cycling

Soil has a crucial function in the global carbon cycle, with the global soil carbon pool estimated to be 2,500 gigatons. This is 3.3 times the amount of the atmospheric pool at 750 gigatons and 4.5 times the biotic pool at 560 gigatons. The pool of organic carbon, which occurs primarily in the form of SOM, accounts for approximately 1,550 gigatons of the total global carbon pool, with soil inorganic carbon (SIC) accounting for the remainder. The pool of organic carbon exists in dynamic equilibrium between gains and losses; soil may therefore serve as either a sink or source of carbon, through sequestration or greenhouse gas emissions, respectively, depending on exogenous factors.

AGROECOSYSTEM MANAGEMENT

Incentivizing a systems approach to agricultural management (rather than a problem-response approach) could support sustainable production as well as ecosystem services such as climate regulation, wildlife conservation, and biological pest control and pollinator management. Note that the ways in which ecosystems produce services are insuffi ciently understood, and that this uncertainty needs to be accounted for in the decision-making process. They advise that future management questions will have to address the complexity of ecosystems in their social context in order that ecological services can be maintained, and also to assess the degree to which technology can substitute for ecological services.

The ecosystem services framework provides a useful umbrella for this endeavour as it can only be achieved by healthy agroecosystems. Sustainable management plans have been advocated for various agroecosystems, ranging from hyper-arid and dryland systems to wetland and aquatic ecosystems. Furthermore, managing agroecosystems for the delivery of multiple services considerably improves the value of the land. For instance, the on-site costs of nutrient depletion (including soil loss through erosion) in the agricultural sector of sub-Saharan Africa vary between countries from less than 1% to more than 20% of the agricultural gross domestic product (GDP). The off-site costs, especially in controlling erosion, can be much larger, and affect a variety of non-agricultural ecosystems and their services. The protection of these services by reducing soil, water and land degradation appears to be a cost-effective investment. Payments for environmental services and other finance mechanisms could be good incentives to use for stopping these off-site costs, but they would be context specific.

Managing Livestock

With their many environmental impacts on soil, water and the atmosphere, there are many opportunities for ecosystem gains in livestock production systems. For instance, the high emission of greenhouse gases can be mitigated by practices such as carbon sequestration in rangelands or improved pastures, by reversing deforestation for the production of feedstuffs through increased agricultural productivity and by using other methods of intensifi cation. Much can also be done by keeping fewer, but more productive, animals by means of better nutrition, animal health, breeding and husbandry techniques. Another innovative approach is the establishment of community-based breeding programmes for the purpose of genetic improvement (e.g. in Ethiopia, where breeding animals are being selected based on phenotypes recorded within the village population). To mitigate greenhouse gas emissions from animal waste, options lie in increased feed digestibility, better storage and treatment of the waste and the appropriate application of waste. There are many other suggestions on how livestock could make a positive contribution to ecosystem services; however, the implementation of some of the proposed alternatives, such as payments for carbon sequestration in rangelands, remains a challenge.

Similarly, health hazards and the pollution of land and water by livestock excreta could be turned around into enhanced nutrient cycling and increased soil water holding capacity by improved management practices. The most effective methods for addressing these problems in catchments are at the farm or production facility. Additional measures can control the effects of manure in watercourses, e.g. manure can be intercepted and stored in ponds, contaminated water can undergo on-farm treatment and constructed farm wetlands can be used to reduce the pathogen load. A potential method described by Masse et al. for developing more sustainable livestock operations utilizes anaerobic digestion biotechnologies to produce biogas, and by this means reduces the need for supplementary chemical nitrogen and phosphorus fertilizers.

The recovery of nutrients from manure, an important contribution to the supporting ecosystem service of nutrient cycling, is highly variable. Approximately 65% of manure nitrogen is recovered from (industrialized) intensive systems in Europe. Almost 30% of this is lost during storage and the maximum cycling efficiency as nitrogen available to crops is around 52%, with large differences between countries. In developing countries too, there is a large range of variation in nitrogen cycling efficiencies in manure management systems. Manure handling and storage, and synchronizing mineralization with crop uptake – and hence fine tuning nutrient cycling in the soil, are key ways in which nitrogen cycling efficiencies can be increased in mixed intensive systems, thus contributing to better regulation of water quality. Results from a recent study in England support earlier conclusions that additions of manure organic carbon produce measureable changes in a wide range of soil biophysical and physicochemical properties and processes that are central to the maintenance of soil fertility and functioning. Smallholder farmers in Africa, who use little fertilizer, recognize the important role of manure in the efficient management and maintenance of soil fertility for crop production. Alternative management of livestock production systems shows that combinations of intensification, better integration of animal manure into crop production and matching the nitrogen and phosphorus supply to livestock requirements can effectively reduce nutrient flows.

With respect to nutrient cycling, therefore, adjustments are needed both in nutrientdeficient systems, where soil fertility is being depleted, and in nutrient-loaded systems, where groundwater contamination, surface water eutrophication and soil pollution are major problems. Technical solutions for reducing the quantity of animal waste and facilitating its proper management and application have to be supported by regulatory measures and financial instruments, such as subsidies and taxes. In nutrient deficient systems, the proper integration of livestock and crop production components in mixed and agropastoral systems can alleviate nutrient export through the application of manure and urine to cultivated areas.

Trees for Agroecosystem Services

A long tradition of separate science and practice in forestry and agriculture means that there are largely untapped opportunities for using trees constructively in agricultural landscapes to sustain food production, while improving a range of ecosystem services. Trees have great potential to play an important role in the sustainable management of agroecosystems. In addition to having impacts on the supporting, regulatory, and cultural services of ecosystems, trees in agroecological landscapes may increase provisioning services by contributing fruit, fodder, fuelwood and timber.

The impact of changing tree cover on various ecosystem services depends on its amount, spatial configuration, species com - position and management. So there is a need to consider planned tree cover change at a landscape scale with the aim of meeting specific suites of objectives, including consideration of the trade-offs and synergies among the ecosystem services affected. The enhancement of tree cover on farmland has the potential to tighten nutrient, water and carbon cycles, and promote the abundance and activity of soil organisms, thereby increasing and sustaining soil and water productivity. Different tree species root to different depths, have leaves at different times throughout the year, and use more or less water through transpiration, attributes that are all affected by management practices such as pruning.

Land Management

A variety of soil conservation techniques are available that can be integrated into agricultural and other land use practices to sustain and enhance agroecosystems and minimize their adverse impacts on their closer environment. Integrated solutions for tackling land degradation can lead to improved water productivity and environmental health, without reducing water availability for food and feed production. An example from Ethiopia describes how successful approaches integrate water and land management with improved agricultural practices, but more examples exist of solutions developed for multifunctional agroecosystems.

Payments for Ecosystem Services

Payments for ecosystem services (PES), also known as payments for environmental services (or benefits), is the practice of compensating individuals or communities for undertaking actions that increase the provision of ecosystem services such as water purification, flood mitigation and carbon sequestration. PES comes under the heading of economic or market-based incentives aimed at motivating the desired decision taking through charges, tradable permits, subsidies and market friction reductions. While the term 'PES' has been in common use since the 1990s, PES type schemes have been around since at least the 1930s when, in the wake of the American Dust Bowl, the federal government paid farmers to avoid farming on poor quality erodible land.

Various studies are discussed in Dunn; these look at the changing drivers for agriculture and at growing urbanization, which both threaten water quality, and at how organizations have set up PES schemes with local farmers. For example, companies pay farmers to adopt less intensive farming techniques, such as outdoor grazing, instead of fertilizer-intensive crop cultivation and feedlots, and the planting of trees to improve soil conditions and promote filtration services. Payments provide sufficient incentives to compensate the famers for these actions, and are developed in collaboration with famers and academics, and negotiated with each farmer. They are intended to reward services that go beyond what is legally required. Such schemes have documented successes in terms of their impacts on water quality, farmer profitability and biodiversity outcomes.

In several of the integrated watershed programmes that have been implemented in India, upstream farmers are compensated for changing their practices, but not necessarily always in cash. Hence, demand for a wide range of ecosystem services from agriculture will increase owing to a greater awareness of both their value and the costs inherent in their depletion.

Today, there are literally hundreds of ongoing PES schemes of all shapes and sizes, all over the world. Some are directed towards achieving poverty reduction on a local level; others maximize the output of goods on an industrial scale. However, all of the schemes essentially involve three steps. First, an assessment of the range of ecosystem services that flow from a particular area, and who they benefit. Secondly, an estimate of the economic value of these benefits to the different groups of people. Finally, a policy, a subsidy or a market to capture this value and compensate individuals or communities for their action. In China's renowned 'Grain for Green' programme, the government thus compensates farmers with grain and cash for planting trees on their sloping farmlands.

Developing mechanisms to implement PES is challenging, not least because although the concept is simple, the reality of making such schemes operational can be very complex, and budgetary resources are often a constraint – especially in poorer countries. Nevertheless, PES can trigger creativity in finding innovative solutions. When effectively designed, PES schemes can give both providers and users of ecosystem services more accurate indications of the consequences of their actions, so that the mix of services provided matches more closely the true preferences of the society concerned. This is the case in Brazil, where water users pay for measures that prevent pollution and erosion. Water users themselves rarely take the initiative but, in Nepal, a fi shing community has developed its own, demand-led, mechanism to ensure good water quality.

A related and comparable concept is that of green water credits, where incentives are given for sound water management or sediment control by appropriate tillage methods or other eco-efficient farming techniques. The idea is to create investment funds so that farmers can take intervention measures for better management of soil and water upstream, which will then be paid for by downstream users that receive more and better quality water.

Ecological Agriculture

Another way of targeting more ecosystem services in agriculture is through alternative approaches to agriculture that are more sustainable and safeguard ecosystem services, in particular from the point of view of water management. Several tools and approaches have been used to implement the concept of sustainable agriculture, such as sustainable land management, ecoagriculture, conservation agriculture, conservation farming, organic agriculture, increased genetic diversity in the production system and others. There are also successful local experiences that have made a paradigm shift away from single solutions to using a portfolio of methods to promote sustainable agriculture; this process should meet the following criteria.

- Ensure that the basic nutritional requirements of present and future generations are met both qualitatively and quantitatively, while providing a number of other agricultural products and ecosystem services.

- Provide durable employment, sufficient income, and decent living and working conditions for all those engaged in agricultural production.
- Maintain and where possible, enhance the productive capacity of the natural resource base as a whole, and the regenerative capacity of renewable resources, without disrupting the functioning of basic ecological cycles and natural balances, or destroying the sociocultural attributes of rural communities, or causing contamination of the environment.
- Reduce the vulnerability of the agricultural sector to adverse natural and socioeconomic factors and other risks, and strengthen self-reliance.

Examples of such Parallel to the demand for more from Kenya: the programme to regain the eroded uplands of Machakos by the Akamba people; and projects carried out by SACDEPKenya (Sustainable Agriculture Community Development Programmes in Kenya).

Conservation agriculture also tries to increase ecosystem services in agriculture, mainly through reducing tillage and restoring land cover, Its primary purpose is to bring water back into the soil and keep it there, but it can have much larger benefits. The Itaipu case demonstrates that, by considering and managing ecosystem functions and services, win–win solutions for both agriculture and other needs can be achieved. The interventions made have increased agricultural productivity and sustainability, in addition to delivering benefits to other ecosystems, such as reduced erosion. The move from conventional agricultural and environmental management practices to nonconventional practices such as conservation agriculture represents a great challenge in terms of changing habits and minds.

Ecoagriculture is another of the many approaches towards sustainable farming, and is highlighted in this book because of its landscape scale and its compatibility with modern high input agriculture. It is 'the design, adaptation and management of agricultural landscapes to produce ecosystem services (e.g. watershed services, wild biodiversity) and generate positive co-benefits for production, biodiversity, and local people, while addressing climate change challenges. Such integrated agricultural landscapes provide critical watershed functions through careful rain and soil water management. This integrated management encompasses the choice of water-conserving crop mixtures, soil and water management (including irrigation), the maintenance of soils to facilitate rainfall infiltration, vegetation barriers to slow the movement of water down slopes, year-round soil cover, and maintenance of natural vegetation in riparian sites, wetlands and other strategic areas of the watershed.

Parallel to the demand for more sustainable agriculture, the health sector has developed interdisciplinary approaches such as 'One Health', striving to attain optimal health for people, animals and our environment, and 'Ecohealth', a participatory methodology for understanding and promoting health and wellbeing in the context of social and ecological interactions. Both of these methods fit well within an ecological approach to agriculture as the two integrated health approaches emphasize a multidisciplinary process and the importance of agriculture and ecosystem-based interventions. This makes them highly suitable for addressing water-related diseases, in a manner that is complementary to that of sustainable agriculture. Agricultural practices that create health risks, such as those related to water management, obviously require farmlevel interventions, and food-borne diseases require management along the 'field-to-fork', or 'boat-to-throat' risk pathway. This includes management of water used at different stages, be it as a production input, in processing, or in meal preparation. Most zoonoses need veterinary and agroecological interventions in addition to medical interventions, as they cannot be controlled as long as diseases remain in the animal reservoir. For zoonoses transmitted through water (e.g. leptospirosis) or via aquatic hosts (e.g. schistosomiasis) interventions may also need to be directed at the aquatic ecosystems.

Farming practice	Conventional farming	Conservation agriculture	Rationale
Tillage	Farmers plough and hoe to improve the soil structure and control weeds	Direct planting without prior inversion of the soil. Planting on the rip line or making holes for planting with a hoe	In the long term, ploughing destroys the soil structure and contributes to declining fertility and levels of organic matter
Crop residues	Farmers remove or burn residues or mix them into the soil with plough or hoe	Crop residue left on the field Planting of cover crops	Crop residues improve soil structure
Mix and rotate crops	Monocultures or crop rotations in a tillage framework where the soil is inverted with a mouldboard plough or similar implement	Crop rotation or intercropping is a permanent feature of the cropping system	Helps to maintain soil fertility Breaks disease cycles

Managing Biological Diversity within Agroecosystems

Recently, more attention has been given to the role of the biological diversity of cultivated ecosystems in providing ecosystem regulating and supporting services. There is a growing body of literature that functional diversity – the value and range of species traits rather than just species numbers – is important to short-term ecosystem resource dynamics and long-term ecosystem stability, as it increases positive interactions or complementary functions. First, crop genetic diversity has been shown to have a direct effect on the maintenance of ecosystem services by providing both: (i) increased numbers of functional traits; and (ii) facilitative interactions that maintain above- and belowground associated biodiversity. This has been shown to be useful in pest and disease management, and has the potential to enhance pollination services and soil processes (nutrient cycling, decomposition and erosion control) in specific situations. Secondly, by increasing long-term stability of the ecosystem in the face of biotic and abiotic stresses and socio-economic variability, crop genetic diversity promotes the continuous maintenance of biomass and the ecosystem services that it provides.

Maintaining or increasing the genetic diversity within the farmer's production system through the use or development of varietal mixtures, or of sets of varieties with nonuniform resistance, has been an alternative agricultural management practice for regulating pests and diseases in many parts of the world. The main purpose of genetic mixtures (crop variety mixtures) for pest and disease management is to slow down the spread of pests and pathogens. Recent studies have shown that a diverse genetic basis of resistance is beneficial for the farmer because it allows a more stable management of pest and disease pressure than does a monoculture. The high levels of diversity of traditional rice varieties in Bhutan have been shown to have high functional diversity against rice blast. Increased levels of common bean and banana diversity in Uganda when disease levels were high showed a significant reduction in pest and disease damage in farmers' fields.

There is growing evidence of the potential of crop genetic diversity to enhance an agroecosystem's capacity to sustain biomass levels through improving the resilience and resistance to environmental variability of that system. High levels of crop genetic diversity occur most commonly in areas where the production environment itself is extremely variable. Here, crop genetic diversity, through its increased portfolio of types, provides the capacity to cope with multiple stresses and changing conditions, thereby ensuring a more stable vegetation cover under a less predictable environment.

The provision of ecosystem services that support soil, water and nutrient availability, and consequently biomass yield, is a management issue that also has the potential to be addressed through crop genetic diversity. In Nepal, farmers typically plant several varieties of rice to match the soil, moisture and other micro-ecological conditions in upland, lowland and swamp environments, which are often all found on the same farm. More than twice the number of rice varieties are found in the hills (which are generally more prone to erosion) as in the lowlands; moreover, farming on slopes tends to be associated with greater diversity in both crops and varieties. In these cases, tolerant varieties are planted where there would otherwise be no vegetative cover, and multiple varieties are planted to best match soil type. This provides for a more continuous planted biomass, and so avoids or decreases soil erosion (and at the same time enhances the soil's ability to sequester carbon).

There are well-documented cases where the low fruit set of crops – and the resulting reduction in yield – has been clearly attributed to pollinator impoverishment. As most temperate and tropical fruit trees are obligatory outcrossers, and rely on insects or small animals for pollination, there is great potential for enhancing the role of the varietal diversity of the fruit trees themselves in promoting cross-hybridization and better fruit production. Studies have shown that strategic plantings, alternating different varieties in a chequerboard pattern for example, can optimize effective pollination visits to two varieties of different attractiveness and at the same time, promote cross-hybridization and better fruit production. In a similar approach, pollinator-attracting genotypes of certain crops have been explored as a management strategy for enhancing pollination services, as genetic polymorphism in the reproductive characters of flowering plants can influence pollinator foraging. Diversity that promotes staggered flowering times among crop varieties has the potential to prolong season-long visitation by bees throughout the protracted flowering season (thus increasing the chances of pollinator population survival to the next growing season), as well as to increase the types of bees visiting at different times during the season, because several bee species are sensitive to climatic variation. In the Yucatan, Mexico, this management strategy is used with maize varieties; short-cycle maize and the more popular long-cycle maize are planted together in order to supply bees with pollen during the wet season and sustain the bee population until the next floral season.

CHALLENGES TO AGROECOSYSTEM MANAGEMENT

The impacts of population growth and other demographic changes on ecosystems can vary over time. Population growth and urban sprawl will result in more people using more resources and placing more pressure on ecosystem services. Increasing populations require more habitable and arable land, which often results in the conversion of natural ecosystems and, ultimately, in the breakdown of ecosystems. There is increasingly negative feedback concerning the interactions between food security, agriculture, water and ecosystem services. Food security is further threatened by reduced yields associated with depleted water quantity, reduced water quality, degradation of other natural resources (such as soil fertility) and the simplification of agricultural systems that have lost their inherent biotic components for regulating pest and disease infestations. Unsustainable agricultural practices can have profound, damaging side effects on livelihoods and ecosystem functioning, and in the long term could potentially depress or reverse productivity gains and increase poverty. At the same time, the availability of other natural resources (land, phosphorus and energy) is predicted to start running out by the end of this century. Efforts to reactivate farmland, e.g. through the use of agrochemicals, have a substantial impact on other ecosystem functions. In turn, dysfunctional ecosystem services further affect the agroecosystems and their production systems.

The Millennium Ecosystem Assessment (MA) of 2005 suggested that in the next 50–100 years, major agricultural decisions would come in the form of trade-offs, especially 'between agricultural production and water quality, land use and biodiversity, water use and aquatic biodiversity, and current water use for agricultural production'. Four scenarios and an adapted version of the MA framework were used in Australia to identify trade-offs between the ecosystem service of water regulation and stakeholders in the Great Barrier Reef's Tully–Murray Catchment. While the most direct trade-off was found to be food and fibre production versus water quality regulation, synergies were also identified with floodplain fisheries.

The greater understanding and appreciation of the role of the services provided by a variety of ecosystems, including agroecosystems, could assist in moving beyond 'trade-offs' to address the challenges of ecosystem management for longterm sustainable food production in many ways. The growing demands for food, coupled with land and water management practices that cause degradation and erode the natural resource base, place substantial constraints on the ecosystem services provided by and inherent within these agroecosystems.

Agriculture and ecosystem services are thus interrelated in at least four ways: (i) agroecosystems generate beneficial ecosystem services such as soil retention, food production and aesthetic benefits; (ii) agroecosystems receive beneficial ecosystem services from other ecosystems, such as pollination from non-agricultural ecosystems; (iii) ecosystem services from non-agricultural systems may be affected by agricultural practices; and finally (iv) the biological diversity within agricultural ecosystems provides regulating and supporting ecosystem services in addition to production services. For food security in the short term, provisioning services are crucial; however, for securing access to food for all in the future, and in the long term, regulatory and supporting services are just as important. The ecosystem services approach requires adaptive management, because its implementation depends on local, national or even global conditions.

Comparing the Economic Values of Ecosystem Services

Decisions on the management of agroecosystem services will typically involve balancing social, economic and environmental considerations, some of them among different services. For example, managing a landscape to maximize food production will probably not maximize water purification for people downstream, and native habitats conserved near agricultural fields may provide both crop pollinators and crop pests. The question about whether intensive or extensive agriculture best optimizes the various trade-offs associated with the provision of ecosystem services is an important issue requiring targeted research.

Connections between ecological sustainability and human well-being can be expressed by using the concept of 'ecological character': the various components and processes in an ecosystem that underpin the delivery of ecosystem services. Without managing for the sustainability of ecological character, the longterm ability of an ecosystem to support human well-being may be compromised. These kinds of management trade-offs often require decision makers to estimate the marginal values of ecosystem services, and to capture the costs and benefits of a specific quantity and quality of services for men and women and different social groups. Marginal value is used in this process because monetary valuation cannot express the overall importance of environmental goods and services, only the value of the resource if there were to be a little more or a little less of it herefore, the value of an ecosystem service reflects its availability. Water is a good example here. It is important and renewable but not replaceable. However, water is often provided freely or at a minimal cost to consumers. The price to consumers only pays for the cost of transmitting water (e.g. water treatment plants), which does not reflect the value of the water itself and gives no information on what consumers would be willing to pay if there were a little more or a little less of the resource.

Ecosystem services can also be used to compare different ecosystem types in terms of their contributions to the availability of a certain service. Most commonly, 'total valuation' is the tool used to bring environmental services into decision-making processes where trade-offs between conservation and development need to be comparatively assessed. Total valuation attempts to account for all of the characteristics of an ecosystem; these include 'its resource stocks or assets, flows of environmental services, and the attributes of the ecosystem as a whole'. This is an incomplete process that is limited in its capacity to value ecosystems fully, though as Daily points out, 'markets play a dominant role in patterns of human behavior, and the expression of value – even if imperfect – in a common currency helps to inform the decision-making process'.

For the quantification of the values of ecosystem services at the country level, a useful concept has been proposed by Dasgupta, who argues that neither gross domestic product (GDP) nor the human development index (HDI) can determine whether development is sustainable. An assessment of wealth per capita is much more useful as it includes the total of all capital assets: infrastructure such as buildings and roads, health, skills, knowledge and institutions, and also natural capital, which may easily be left out of other assessments.

These methods are increasingly important to today's decisions on agricultural water use. Bennett et al. point out that, with growing demands on food production and water use, demands on ecosystem services, in many cases, could surpass the capacity of certain ecosystems to supply these services. In these contexts, decision makers will need to draw a balance between the production of various services in ecosystems on the one hand, and the social and economic benefits and risks of using technology to provide them on the other. With a clear understanding of ecosystem services and their values, agroecosystems and nonagricultural terrestrial ecosystems can be compared. Many goods and provisioning services come from nonagricultural land (such as food, fodder, fibre and timber), and in decisions over water allocation the whole range of ecosystem services, their benefits (values) and costs (social, financial, water) have to be taken into account. Only then can well-balanced decisions be made about which ecosystem services are to be enhanced, at the expense of which other services, or about how ecosystems can be optimized to provide the widest range of ecosystem services.

Finally, any successful decision making will depend on farmers and the farming community having the knowledge and leadership capacity to evaluate the benefits that any action will have for them. This, in turn, will be dependent on systems that are in place that support activities taken by local, national and international organizations and agencies towards strengthening local institutions so as to enable farmers to take a greater role in the management of their resources.

Agroecosystems have an important role to play in food security but they have also been associated with negative impacts on other ecosystems. When compared with other groups of ecosystems, or biomes, the total value of ecosystem services from cropland is relatively low, even for food production alone. For example, in the Mississippi Delta, the total annual value of agricultural land ranged from US $195 to US $220/ha, of which US $85/ha was from food production; this level of production fell behind that of most other ecosystem types (including forests and in particular, wetlands, where annual food production was valued at US $145 to US $3346/ha).

In contrast, some other studies have found higher annual values for food production in cultivated systems: US $667/ha in South Africa, US $1516/ha in El Salvador, and as high as US $3842/ha and US $7425/ha in Israel. There have been very few studies that have attempted to value ecosystem services in agriculture, even though assessments indicate that the value to agriculture is enormous, and various estimates do suggest a real underestimation of the benefits of non-agricultural ecosystems for food production and possibly for food security. In a study in Denmark, Porter et al. estimated via field scale ecological monitoring and economic value-transfer methods, the market and non-market ecosystem service value of a combined food and energy (CFE) agroecosystem that simultaneously produces food, fodder and bioenergy.

Discrepancies in estimations of the economic values of ecosystem services occur, in part, because land and water use planning are based on limited sector-based considerations, which do not factor in the overall values of all services that any ecosystem delivers. Hence, agricultural land has such a low value in terms of output because it tends to be managed for a single service (food production), often with significant negative consequences on other services (e.g. through pollution). Another reason might be that the value of food production is measured in terms of market prices, whereas the value of other ecosystem services reflects avoided societal costs that are normally much higher but for which there are no marketplaces (with the exception of carbon). Nevertheless, food production will always remain a priority and does not necessarily have to come at the expense of other services. Cases exist in which investments in sustainable agriculture have generated co-benefits in raising food production, while at the same time improving ecosystem services and functions.

Understanding Agroecosystem Services

Managing agricultural land to deliver multiple services considerably improves the value of the land. However, in order to enhance improved services – such as carbon storage, erosion control, water retention, waste treatment, regulation of pests and diseases, and cultural and recreational values including tourism – their values must be understood in comparison with agricultural income. Ideally, these added services would not conflict with agricultural production in many cases but rather improve both its productivity and its sustainability, with beneficial impacts on surrounding ecosystems as well.

Over the years, agricultural systems have evolved into diverse agroecosystems, some of which are rich in biodiversity and provide ecosystem services in addition to food production. Examples are wet rice–poultry farming systems and the practice of increased diversity of crop varieties within farmers' fields which has been shown to reduce the risk of crop loss to pest diseases.

Water management in agroecosystems can create competition with wider environmental requirements and affect water flow downstream. Decisions on water use require mechanisms in which the needs of both the farmers and the ecosystem services are met, e.g. by buying irrigation water from farmers to sustain or rehabilitate ecosystems and their services These decisions need a broader consideration of ecosystem services in agroecosystems. This consideration should take into account which services are enhanced at the expense of which other ecosystem services, and which services benefit mostly poor women, men and other vulnerable groups. In agroecosystems, food production is again underpinned by a reliable availability of water. Tools, such as the polyscape tool, are being developed that allow the quantification of trade-offs and synergies among the impacts of water- and land-use interventions on different ecosystem services.

Agriculture is thus faced with significant challenges regarding water use and availability. Solutions, which are based largely on the more efficient use of water in agriculture, do exist but agriculture can also be managed differently, in such a way as to enhance ecosystem services and increase the capacities of low-income male and female farmers. This change in thinking, and in the way that agroecosystems are managed, is crucial for global food security. The major challenge then lies in quantifying values and measuring feedback cycles and more research is required into ecosystem services, especially those associated with water and with other components of the agricultural production systems.

Ecosystem Services and Fish

Important fisheries that depend on healthy aquatic ecosystems are endangered. Because fish provide 21% of animal protein in Africa, and 28% in Asia, a loss of fisheries can be detrimental for food security. The link with management of inland aquatic ecosystems is clear, as almost 50% of global fish consumption comes from aquaculture, and in Africa almost half is from inland fisheries. In order to avoid further degradation, fundamental changes are required to establish an eco system based catchment management approach.

Faced with declining wild fish stocks, overexploitation of target species and by-catch of other species, the fishing industry is giving way to aquaculture, which is reported to be the world's fastest growing food sector – at an average growth rate of 6.8%/year. In 2006, it was reported that 53 million t of fish (or half of all fish consumed in the world) were produced by the aquaculture industries. At the same time, fisheries are increasingly less 'wild', as stock enhancement and the establishment of culture-based fisheries are increasingly viewed as potential means of bolstering catches. None the less, the potential negative ecological and social impacts of such practices demand comprehensive and rigorous assessment, with appropriate mitigation and control measures, before they are implemented. For example, antibiotics and other chemicals used in fi sh farms can seep into surrounding waters, and sensitive coastal areas and wetlands are also disrupted or destroyed in the development of the industry. Additionally, aquaculture appropriates a range of environmental goods and services that may lead to adverse environmental impacts, and affect the ability of stocks and flows of ecosystem services to sustain other productive activities, which could again result in disputes and conflicts.

Ecosystem Impacts of Livestock Production

Livestock systems occupy about 30% to 45% of the planet's ice-free terrestrial surface area. This makes livestock the single largest agricultural use of land globally, either directly through grazing or indirectly through the consumption of fodder and feed grains. Livestock is also a signifi cant global asset, with a value of at least US $1.4 trillion in the least developed countries, excluding the value of infrastructure and land. The accelerating demand for livestock products is increasingly being met by intensive (industrialized) production systems, especially for chickens and pigs in Asia. Thus, between 1995 and 2005, bovine and ovine meat production increased by about 40%, pig meat production rose by nearly 60% and poultry meat production doubled. Livestock production has important implications for ecosystem services, with environmental impacts on water scarcity, nutrient cycling, climate change and land degradation, as well as human impacts such as public health and the exclusion of smallholder producers.

Livestock production emits large amounts of greenhouse gases. However, the mitigation potential in the livestock sector is very large, with improved feeding practices, manure and land use management practices representing over 80% of this potential.

A well-known linkage between livestock and soil productivity is the cycling of biomass (natural vegetation, crop residues) through animals (cattle, sheep, goats) into excreta (manure, urine) that fertilize the soil. Globally, manure contributes 14% of nitrogen, 25% of phosphorus and 40% of potassium nutrient inputs to agricultural soils. The types and amounts of manure nutrients available for recycling are highly influenced by differences in land use and in the spatial and temporal distribution of livestock as dictated by animal management, and also by seasonal differences in animal diet. When not carefully managed, nutrient surpluses from livestock waste and fertilizer used for feed production may result in eutrophication of surface waters and groundwater contamination in places where large animals congregate, such as in industrial periurban systems. These ecosystem disservices posed by water contamination from livestock excreta and dung residues can cause health hazards.

Then again, there are numerous potential situations where co-benefits emerge between livestock production and the maintenance of ecosystem services. These examples may not be readily available, as they require in-depth analysis of scientific as well as indigenous evidence, and therefore come at a (knowledge-intensive) cost. Herrero et al. formulated useful guiding questions on livestock, ecosystems and livelihoods to help to identify knowledge gaps. In analysing environmental impacts and ecosystem services, it is important to distinguish between extensive and intensive livestock production. Although livestock grazing is the largest user of land globally, most of the world's animal production comes from intensive industrialized production in developed countries, closely followed by rainfed mixed crop–livestock systems in developing countries. These intensively farmed areas are the focal points for ecosystem degradation. For example, in Ethiopia, 45% of the estimated soil loss occurs from the 13% of the country under cultivation, but grazing lands, which cover about half of the country, account for only 21% of the soil loss Some livestock herding systems in Africa have managed large areas in a seminatural state, maintaining vegetation cover and indirectly preserving vital ecosystem services.

Sustainable growth and intensification of livestock production systems will be required to cater for increasing demands for livestock products, while mitigating the negative effects of the sector. Substantive investments and policies are essential to implement the measures above. With more sustainable livestock production systems, the increased demands for animal products could be satisfied at the same time as maintaining environmental flows and services.

Land Degradation and Erosion

Soil degradation, such as by water or wind erosion, compaction, salinization, nutrient depletion and fertility decline, physical deterioration, contamination and sealing, is considered to be a main cause of hampering growth in agricultural productivity. The impact of soil degradation on yields in China was estimated as a reduction in food production capacity on the current arable land area from 482 Mt in 2005 to 412 Mt by 2050, with the same relative yield loss projected in the next 15 years as in the past 15 years, though such estimates do not account for underlying processes; hence, for identifying viable solutions, more detailed studies at a lower level will be required. In addition to physical factors, land degradation has many social roots, including lack of land tenure, careless extractivism, indifferent or corrupt governments, lack of access to finance and resources, population pressure and a dearth of educational opportunities.

In many parts of the world, land degradation has increased over the past two decades, mostly as a result of poor land management, including uncontrolled soil erosion, overgrazing, and the limited application and availability of appropriate types of fertilizers. In sub-Saharan Africa, more than 40% of the land is threatened by land degradation. Loss of organic matter, e.g. through entire crop removal, and the physical degradation of soil not only reduce nutrient availability but also result in lower water infiltration rates and porosity, and these may affect the resilience of agroecosystems, local and regional water productivity, and even global carbon cycles. Accelerated on-farm soil erosion leads to substantial yield losses and contributes to downstream sedimentation, which can degrade natural water bodies and fill up water storage reservoirs and irrigation infrastructure.

The occurrence of land degradation is thus linked with low water productivity and impaired ecosystem services, and is often associated with high population pressure; nevertheless, its extent and its causative mechanisms are highly site specific. One way of dealing with this is to facilitate outmigration of people from vulnerable areas through the provision of education and credit services offering alternative livelihoods. However, high population pressure and market demand can in itself trigger investments in labour-intensive conservation practices and natural resources management.

Another argument for taking a landscape approach, is the role of trees. Recent assessments suggest that almost half of all agricultural land has more than 10% tree cover, indicating that trees are a mainstream component of agricultural landscapes and may provide forest functions to some extent. Tree cover in farming landscapes can have a large impact on the infi ltration and penetration of water and thereby, on catchment hydrology. Furthermore, when tree cover is changed, other ecosystem services besides water flow may also be affected, such as pollination and carbon storage, and these can also infl uence agricultural productivity. The impact of changing tree cover on various ecosystem services depends on its amount, spatial configuration, species composition and management, so there is a need to get beyond generalizations and look at tree cover at the landscape scale in order to meet specific objectives, including the consideration of tradeoffs and synergies among the ecosystem services affected.

References

- *Bland, W.L. and Bell, M.M., (2007) A holon approach to agroecology International Journal of Agricultural Sustainability 5(4), 280-294. Abstract available here Archived April 26, 2012, at the Wayback Machine
- Agricultural-ecosystem, earth-and-planetary-sciences, topics: sciencedirect.com, Retrieved 22 February, 2019
- Ohkuma, M. (2003). "Termite symbiotic systems: Efficient bio-recycling of lignocellulose". Applied Microbiology and Biotechnology. 61 (1): 1–9. Doi:10.1007/s00253-002-1189-z. PMID 12658509
- Biodiversity-for-food-and-agriculture: redock.org, Retrieved 23 March, 2019
- Levin, Simon A; Carpenter, Stephen R; Godfray, Charles J; Kinzig, Ann P; Loreau, Michel; Losos, Jonathan B; Walker, Brian; Wilcove, David S (27 July 2009). The Princeton Guide to Ecology. Princeton University Press. P. 330. ISBN 978-0-691-12839-9
- Soil-Microorganisms, The-Living-Soil, Text-Version, Building-Soils-for-Better-Crops-3rd-Edition, Books, Learning-Center: sare.org, Retrieved 24 April, 2019
- "The Problem of Increasing Human Energy" by Nikola Tesla". Archived from the original on 6 June 2011. Retrieved 22 June 2016

3

Stresses in Agroecology

Stresses in agroecology are categorized as biotic and abiotic stresses. Some of the varied aspects of agroecology stress include interaction of plant biotic and abiotic stress as well as use of agriculture biotechnology for management of these stresses. This chapter discusses in detail these environmental stresses related to agroecology.

Plant stress is a state where a plant is growing in non-ideal growth conditions and has increased demands put on it. Plant stress refers to any unfavorable condition or substance that affects a plant's metabolism, reproduction, root development, or growth. Plant stress can come in different forms and durations. Some plant stressors are naturally occurring, like drought or wind, while others may be the result of human activity, like over irrigation or root disturbance.

Causes of Plant Stress

Plant stress is caused by a variety of factors, some of which are obvious (like natural disasters), while others occur on a micro scale in the soil. Recent natural disasters represent one type of plant stress factors, called abiotic factors, which usually occur above ground. A second type of plant stress factors are called biotic factors, which mostly occur underground, and can cause plant stress through pathogens and pests.

Reducing Plant Stress

Plant stress can be prevented and limited by maintaining a stable soil foundation and optimizing key soil performance characteristics. In farming, gardening, landscaping, and nursery management, soil serves as the foundation for plant quality, beauty, and productivity. So, by optimizing soil health, plants are better able to withstand abiotic and biotic factors.

Optimizing Soil Health

Soil health can be optimized by implementing management practices that maintain a suitable habitat for the soil microorganisms and that build lasting soil structure. These management practices include disturbing the soil as little as possible (consider reduced or no-till disturbance), growing multiple species of plants, keeping living plants in the soil often, and covering the soil with cover crops. Additionally, amendments can be added to the soil to enhance specific soil properties.

Soil Health Practices to benefit Plants and the Planet

Abiotic and biotic plant stress factors will always exist, and climate change may even cause greater plant stress factors in the future. report by the UN Intergovernmental Panel on Climate Change warned that there is only a dozen years for global warming to be kept to a maximum of 1.5 degrees Celsius, beyond which a slight increase will significantly increase the risks of drought, floods, extreme heat, fires, and poverty for millions of people around the world.

The farmers and land stewards most often bear the brunt of the impact of climate change, but fortunately they are also innovating and implementing soil management practices to enable plants to better withstand abiotic and biotic plant stress factors. On top of that, improving soil health is recognized as a method for mitigating the impacts of climate change.

BIOTIC STRESS

Biotic stress is stress that occurs as a result of damage done to an organism by other living organisms, such as bacteria, viruses, fungi, parasites, beneficial and harmful insects, weeds, and cultivated or native plants. It is different from abiotic stress, which is the negative impact of non-living factors on the organisms such as temperature, sunlight, wind, salinity, flooding and drought. The types of biotic stresses imposed on an organism depend the climate where it lives as well as the species' ability to resist particular stresses. Biotic stress remains a broadly defined term and those who study it face many challenges, such as the greater difficulty in controlling biotic stresses in an experimental context compared to abiotic stress.

The damage caused by these various living and nonliving agents can appear very similar. Even with close observation, accurate diagnosis can be difficult. For example, browning of leaves on an oak tree caused by drought stress may appear similar to leaf browning caused by oak wilt, a serious vascular disease caused by a fungus, or the browning caused by anthracnose, a fairly minor leaf disease.

Biotic stressors are a major focus of agricultural research, due to the vast economic losses caused to cash crops. The relationship between biotic stress and plant yield affects economic decisions as well as practical development. The impact of biotic injury on crop yield impacts population dynamics, plant-stressor coevolution, and ecosystem nutrient cycling.

Biotic stress also impacts horticultural plant health and natural habitats ecology. It also has dramatic changes in the host recipient. Plants are exposed to many stress factors, such as drought, high salinity or pathogens, which reduce the yield of the cultivated plants or affect the quality of the harvested products. Although there are many kinds of biotic stress, the majority of plant diseases are caused by fungi. Arabidopsis thaliana is often used as a model plant to study the responses of plants to different sources of stress.

Effect on Plant Growth

Photosynthesis

Many biotic stresses affect photosynthesis, as chewing insects reduce leaf area and virus infections reduce the rate of photosynthesis per leaf area. Vascular-wilt fungi compromise the water transport and photosynthesis by inducing stomatal closure.

Response to Stress

Plants have co-evolved with their parasites for several hundred million years. This co-evolutionary process has resulted in the selection of a wide range of plant defences against microbial pathogens and herbivorous pests which act to minimise frequency and impact of attack. These defences include both physical and chemical adaptations, which may either be expressed constitutively, or in many cases, are activated only in response to attack. For example, utilization of high metal ion concentrations derived from the soil allow plants to reduce the harmful effects of biotic stressors (pathogens and herbivores); meanwhile preventing the infliction of severe metal toxicity by way of safeguarding metal ion distribution throughout the plant with protective physiological pathways. Such induced resistance provides a mechanism whereby the costs of defence are avoided until defense is beneficial to the plant. At the same time, successful pests and pathogens have evolved mechanisms to overcome both constitutive and induced resistance in their particular host species. In order to fully understand and manipulate plant biotic stress resistance, we require a detailed knowledge of these interactions at a wide range of scales, from the molecular to the community level.

Inducible Defense Responses to Insect Herbivores

In order for a plant to defend itself against biotic stress, it must be able to differentiate between an abiotic and biotic stress. A plants response to herbivores starts with the recognition of certain chemicals that are abundant in the saliva of the herbivores. These compounds that trigger a response in plants are known as elicitors or herbivore-associated molecular patterns (HAMPs). These HAMPs trigger signalling pathways throughout the plant, initiating its defence mechanism and allowing the plant to minimise damage to other regions. These HAMPs trigger signalling pathways throughout the plant, initiating its defence mechanism and allowing the plant to minimise damage to other regions. Phloem feeders, like aphids, do not cause a great deal of mechanical damage to plants, but they are still regarded as pests and can seriously harm crop yields. Plants have developed a defence mechanism using salicylic acid pathway, which is also used in infection stress, when defending itself against phloem feeders. Plants perform a more direct attack on an insects digestive system. The plants do this using proteinase inhibitors. These proteinase inhibitors prevent protein digestion and once in the digestive system of an insect, they bind tightly and specifically to the active site of protein hydrolysing enzymes such as trypsin and chymotrypsin. This mechanism is most likely to have evolved in plants when dealing with insect attack.

Plants detect elicitors in the insects saliva. Once detected, a signal transduction network is activated. The presence of an elicitor causes an influx of Ca^{2+} ions to be released in to the cytosol. This increase in cytosolic concentration activates target proteins such as Calmodulin and other binding proteins. Downstream targets, such as phosphorylation and transcriptional activation of stimulus specific responses, are turned on by Ca^{2+} dependent protein kinases. In Arabidopsis, over expression of the IQD1 calmodulin-binding transcriptional regulator leads to inhibitor of herbivore activity. The role of calcium ions in this signal transduction network is therefore important.

Calcium Ions also play a large role in activating a plants defensive response. When fatty acid amides are present in insect saliva, the mitogen-activated protein kinases (MAPKs) are activated. These genes when activated, play a role in the jasmonic acid pathway. The jasmonic acid pathway is also referred to as the Octadecanoid pathway. This pathway is vital for the activation of defence genes in plants. The production of jasmonic acid, a phytohormone, is a result of the pathway. In an experiment using virus-induced gene silencing of two calcium-dependent protein kinases (CDPKs) in a wild tobacco (*Nicotiana attenuata*), it was discovered that the longer herbivory continued the higher the accumulation of jasmonic acid in wild-type plants and in silenced plants, the production of more defence metabolites was seen as well as the decrease in the growth rate of the herbivore used, the tobacco hornworm (*Manduca sexta*). This example demonstrates the importance of MAP kinases in plant defence regulation.

Inducible Defense Responses to Pathogens

Plants are capable of detecting invaders through the recognition of non-self signals despite the lack of a circulatory or immune system like those found in animals. Often a plant's first line of defense against microbes occurs at the plant cell surface and involves the detection of microorganism-associated molecular patterns (MAMPs). MAMPs include nucleic acids common to viruses and endotoxins on bacterial cell membranes which can be detected by specialized pattern-recognition receptors. Another method of detection involves the use of plant immune receptors to detect effector molecules released into plant cells by pathogens. Detection of these signals in infected cells leads to an activation of effector-triggered immunity (ETI), a type of innate immune response.

Both the pattern recognition immunity (PTI) and effector-triggered immunity (ETI) result from the upregulation of multiple defense mechanisms including defensive chemical signaling compounds. An increase in the production of salicylic acid (SA) has been shown to be induced by pathogenic infection. The increase in SA results in the production of pathogenesis related (PR) genes which ultimately increase plant resistance to biotrophic and hemibiotrophic pathogens. Increases in jasmonic acid (JA) synthesis near the sites of pathogen infection have also been described. This physiological response to increase JA production has been implicated in the ubiquitination of jasmonate ZIM domains (JAZ) proteins, which inhibit JA signaling, leading to their degradation and a subsequent increase in JA activated defense genes.

Studies regarding the upregulation of defensive chemicals have confirmed the role of SA and JA in pathogen defense. In studies utilizing Arabidopsis mutants with the bacterial NahG gene, which inhibits the production and accumulation of SA, were shown to be more susceptible to pathogens than the wild-type plants. This was thought to result from the inability to produce critical defensive mechanisms including increased PR gene expression. Other studies conducted by injecting tobacco plants and Arabidopsis with salicylic acid resulted in higher resistance of infection by the alfalfa and tobacco mosaic viruses, indicating a role for SA biosynthesis in reducing viral replication. Additionally, studies performed using Arabidopsis with mutated jasmonic acid biosynthesis pathways have shown JA mutants to be at an increased risk of infection by soil pathogens.

Along with SA and JA, other defensive chemicals have been implicated in plant viral pathogen defenses including abscisic acid (ABA), gibberellic acid (GA), auxin, and peptide hormones. The use of hormones and innate immunity presents parallels between animal and plant defenses, though pattern-triggered immunity is thought to have arisen independently in each.

Cross Tolerance with Abiotic Stress

- Evidence shows that a plant undergoing multiple stresses, both abiotic and biotic (usually pathogen or herbivore attack), can produce a positive effect on plant performance, by reducing their susceptibility to biotic stress compared to how they respond to individual stresses. The interaction leads to a crosstalk between their respective hormone signalling pathways which will either induce or antagonize another restructuring genes machinery to increase tolerance of defense reactions.

- Reactive oxygen species (ROS) are key signalling molecules produced in response to biotic and abiotic stress cross tolerance. ROS are produced in response to biotic stresses during the oxidative burst.
- Dual stress imposed by ozone (O_3) and pathogen affects tolerance of crop and leads to altered host pathogen interaction. Alteration in pathogenesis potential of pest due to O_3 exposure is of ecological and economical importance.
- Tolerance to both biotic and abiotic stresses has been achieved. In maize, breeding programmes have led to plants which are tolerant to drought and have additional resistance to the parasitic weed Striga hermonthica.

Remote Sensing

The Agricultural Research Service (ARS) and various government agencies and private institutions have provided a great deal of fundamental information relating spectral reflectance and thermal emittance properties of soils and crops to their agronomic and biophysical characteristics. This knowledge has facilitated the development and use of various remote sensing methods for non-destructive monitoring of plant growth and development and for the detection of many environmental stresses that limit plant productivity. Coupled with rapid advances in computing and position locating technologies, remote sensing from ground-, air-, and space-based platforms is now capable of providing detailed spatial and temporal information on plant response to their local environment that is needed for site specific agricultural management approaches. This is very important in today's society because with increasing pressure on global food productivity due to population increase, result in a demand for stress-tolerant crop varieties that has never been greater.

ABIOTIC STRESS

Abiotic stress is stress produced by natural environment factors such as extreme temperatures, wind, drought, and salinity. Humankind doesn't have much control over abiotic stresses. It is very important for humans to understand how stress factors affect plants and other living things so that we can take some preventative measures.

Preventative measures are the only way that humans can protect themselves and their possessions from abiotic stress. There are many different types of abiotic stressors, and several methods that humans can use to reduce the negative effects of stress on living things.

Cold

One of the types of Abiotic Stress is cold. This has a huge impact on farmers. Cold impacts crop growers all over the world in every single country. Yields suffer and farmers also suffer huge losses because the weather is just too cold to produce crops.

Humans have planned the planting of our crops around the seasons. Even though the seasons are fairly predictable, there are always unexpected storms, heat waves, or cold snaps that can ruin our growing seasons.

ROS stands for reactive oxygen species. ROS plays a large role in mediating events through transduction. Cold stress was shown to enhance the transcript, protein, and activity of different ROS-scavenging enzymes. Low temperature stress has also been shown to increase the H_2O_2 accumulation in cells.

Plants can be acclimated to low or even freezing temperatures. If a plant can go through a mild cold spell this activates the cold-responsive genes in the plant. Then if the temperature drops again, the genes will have conditioned the plant to cope with the low temperature. Even below freezing temperatures can be survived if the proper genes are activated.

Heat

Heat stress has been shown to cause problems in mitochondrial functions and can result in oxidative damage. Activators of heat stress receptors and defenses are thought to be related to ROS. Heat is another thing that plants can deal with if they have the proper pretreatment. This means that if the temperature gradually warms up the plants are going to be better able to cope with the change. A sudden long temperature increase could cause damage to the plant because their cells and receptors haven't had enough time to prepare for a major temperature change.

Heat stress can also have a detrimental effect on plant reproduction. Temperatures 10 degrees Celsius or more above normal growing temperatures can have a bad effect on several plant reproductive functions. Pollen meiosis, pollen germination, ovule development, ovule viability, development of the embryo, and seedling growth are all aspects of plant reproduction that are affected by heat.

There have been many studies on the effects of heat on plant reproduction. One study on plants was conducted on Canola plants at 28 degrees Celsius, the result was decreased plant size, but the plants were still fertile. Another experiment was conducted on Canola plants at 32 degrees Celsius, this resulted in the production of sterile plants. Plants seem to be more easily damaged by extreme temperatures during the late flower to early seed development stage.

Wind

Wind is a huge part of abiotic stress. There is simply no way to stop the wind from blowing. This is definitely a bigger problem in some parts of the world than in others. Barren areas such as deserts are very susceptible to natural wind erosion. These types of areas don't have any vegetation to hold the soil particles in place. Once the wind starts to blow the soil around, there is nothing to stop the process. The only chance for the soil to stay in place is if the wind doesn't blow. This is usually not an option.

Plant growth in windblown areas is very limited. Because the soil is constantly moving, there is no opportunity for plants to develop a root system. Soil that blows a lot usually is very dry also. This leaves little nutrients to promote plant growth.

Farmland is typically very susceptible to wind erosion. Most farmers do not plant cover crops during the seasons when their main crops are not in the fields. They simply leave the ground open and uncovered. When the soil is dry, the top layer becomes similar to powder. When the wind blows, the powdery top layer of the farmland is picked up and carried for miles. This is the exact scenario that occurred during the "Dust Bowl" in the 30's. The combination of drought and poor farming practices allowed the wind to moves thousands of tons of dirt from one area to the next.

Wind is one of the factors that humans can really have some control over. Simply practice good farming practices. Don't leave ground bare and without any type of vegetation. During dry seasons it is especially important to have the land covered because dry soil moves much easier than wet soil in the wind.

When soil is not blowing due to the wind, conditions are much better for plant growth. Plants cannot grow in a soil that is constantly blowing. Their root systems do not have time to be established. Also, when soil particles are blowing they wear away at the plants that they run into. Plants are essentially "sand blasted."

Drought

Drought is very detrimental to all types of plant growth. When there is no water in the soil there are not very many nutrients to support plant growth. Drought also enhances the effects of wind. When drought occurs the soil becomes very dry and light. The wind picks up this dry dirt and carries it away. This action severely degrades the soil and creates a poor condition for growing plants.

Adaptation of Plants

Plants have been exposed to the elements for thousands of years. During this time they have evolved in order to lessen the effects of abiotic stress. Signal transduction is the mechanism in plants that is responsible for the adaptation of plants. Many signaling transduction networks have been discovered and studied in microbial and animal systems. There is limited knowledge in the plant field because it is very difficult to find exactly which phenotypes in the plant are affected by stressors. These phenotypes are very valuable to the researchers. They need to know the phenotypes so that they can create a method to screen for mutant genes. Mutants are the key to finding signaling pathways in living creatures.

Animals and microbes easier to run tests on because they show a reaction fairly quickly when a stress factor is put on them, this leads to the isolation of the specific gene. There have been decades of research on the effects of temperature, drought, and salinity, but not very many answers.

Receptors

The part of the plants, animal, or microbe that first senses an abiotic stress factor is a receptor. Once a signal is picked up by a receptor, signals are transmitted intercellularly and then they activate nuclear transcription to get the effects of a certain set of genes. These activated genes allow the plant to respond to the stress that it is experiencing. Even though none of the receptors for cold, drought, salinity or the stress hormone abscisic acid in plants is known for sure, the knowledge that we have today shows that receptor-like protein kinases, two-component histidine kinases, as well as G-protein receptors may be the possible sensors of these different signals.

Receptor like kinases can be found in plants as well as animals. There are many more RLKs in plants than there are in animals. They are also a little bit different. Unlike animal RLKs that usually possess tyrosine signature sequences, plant RLKs have serine or threonine signature sequences.

Genetically Modified Plants

Plants are most commonly modified to be resistant to specific herbicides or pathogens, but we have the technology to modify plants in order to make them resistant to specific abiotic stressors. Cold, heat, drought, or salt are all factors that could possibly be defended against by genetically modified plants.

Some plants could have genes added to them from other species of plants that have a resistance to a specific stress. Plants implanted with these genes would then become transgenic plants because they have the genes from another species of plant in them. Scientists first have to isolate the specific gene in a plant that is responsible for its resistance. The gene would then be taken out of the plant and put into another plant. The plant that is injected with the new resistant gene would have a resistance to an abiotic stressor and be able to tolerate a wider range of conditions.

This process of creating transgenic plants could have a huge impact on our nation's economy. If plants could be genetically engineered to be resistant to a wider variety of stress, crop yields would skyrocket. With the expansion of town and cities there is a decreasing number of farm acres. Although the farm acres are being built on, the number of people consuming agriculture products is going up. Ethanol is also responsible for using much more of the corn that is grown.

The production of this fuel has put a strain on the corn market. Prices of corn have gone up and this price is having a negative impact on the people who feed animals using corn. The combination of reduced acres of farmland and a higher demand on crops have left producers and consumers in a severe dilemma. The only solution to this problem is to keep getting higher and higher yields from the cropland that we have left.

Genetically modified plants are a good answer to the problem of not enough crops to go around. These plants can be engineered to be resistant to all types of abiotic stress. This would eliminate crop yield loss due to extreme temperatures, drought, wind, or salinity. The consumers of crops would enjoy a little bit lower prices because the demand on them would be a little lower.

The Midwestern U.S. is experiencing a severe drought. Farmers are being limited on how much they can irrigate due to the shortage of water. There is also very little rain during the growing season so the crops do not yield very well. This problem could be solved by genetically modifying plants to become more drought resistant. If plants could use less water and produce yields that are superior or equal to current ones, it would be better for the people and also the environment. People would enjoy an abundance of crops to consume and export for a profit. The environment would be able to have more water in its aquifers and rivers throughout the country.

Another environmental factor that would be improved would be the amount of land left for wildlife. Crops modified to be resistant to abiotic stress and other factors that decrease yields would require less land use. Producers would be able to grow enough crops on less acres if the plants were modified to produce very high yields. This would allow some of the cropland that is in use today to be set aside for wildlife. Instead of farming "fence line to fence line" farmers would be able to create large buffers in their fields. These buffers would provide a great habitat for plants and animals.

A lot of people do not like genetically modified organisms. People opposed to these modified plants often claim that they are not safe for the environment or for human consumption. There are many videos and reports in circulation that discredit the safety of genetically modified organisms. King Corn is one video that claims that corn is bad for humans to consume.

There are strict regulations and protocols that go along with genetically modifying plants. A company that specializes in producing genetically modifying organisms must put their plants through a huge variety of tests to ensure the safety of their product. Each of these tests must be passed by the product in order to produce more of the plant seeds.

When seeds are mass-produced, the fields that they are grown in have to meet specific criteria. They must have no vegetation zones around them to prevent the spread of the modified plants into the native population. The plots must be carefully labeled and marked so that the company knows exactly what is planted in the field. All of these protocols are in place to ensure the safety of the consumers and also of the environment. Because genetically modified plants are given stress resistant genes or high yielding genes they are better for the environment. They only help create more land to be put back into natural habitats for plants and animals.

Influence of Rhizobacteria in Plant Tolerance to Abiotic Stress

The unpredictability of the hydrological cycle has posed a serious challenge to farmers, horticulturists, and to the global community, concerning its effect in meeting food needs of mankind and animals. The number of people to be fed is constantly increasing and food supplies are not meeting the demand.

In order to increase the quantity and quality of crops grown, agriculturists have intensified the use of open and ground water sources for irrigation purposes, which has a corresponding salinization implication.

However, the use of bioinoculants (plant growth-promoting rhizobacteria) has been of great help in combating this abiotic-climate-induced change that limits the overall performance of plants under stress. The use of biofertilizers to enhance successful adaptation and survival of plants will gear toward ensuring a sustainable crop yield and improvement of soil fertility and structure. This is a good approach to stress management. These microbes through deaminase enzyme production, nodulation, and other physiological activities at the rhizosphere help the plants tolerate stress.

External (exogenous) introduction of beneficial chemical substances as supplement has been used to improve resilience, yield, and tolerance of plants to the toxicity of these stress-imposed conditions, such as the application of caffeic acid, sodium polyacrylate, jasmonic acid, chitosan, humic acid, gibberellic acid and other hormones, and molybdenum. These chemicals aid the physiological performance and adaptation that manifest in a well-balanced morphological state of the plant growing in a stressful environment. The stresses range from drought, salinity, alkalinity, pathogenic microbial attack to chemicals.

Abiotic stresses are stress conditions to plants arising from the environment. They are the non-living part of the ecosystem whose effect is felt by the living component of the system. Nature is balance sensitive, and at short supply or deviation from the normal occurrence of these conditions will create a stress in the ecosystem and jeopardize the well-being of living things. Water, nutrients, salts, temperature, and pH are among the basic abiotic components of the agricultural ecosystem that have an influence on plants.

These abiotic stresses not only affect plants but also affect microbes. In an effort to survive and prevail despite the stiff competition among microbes particularly at the rhizosphere, some microbes possessing the cellulase enzyme capable of dissolving the cellulose cell wall of plant roots gain entrance into the apoplast of plants, the cell wall interior as well as the vascular bundle—xylem, where they live and undergo normal metabolic activities.

Although researchers have looked at various means of handling the stress inducers to enable plants to increase their survival and performance, yet, little has been done on the integrative aspect of this enhancement strategy of protecting and empowering plants to resist and grow better under drought, salinity, or alkaline conditions.

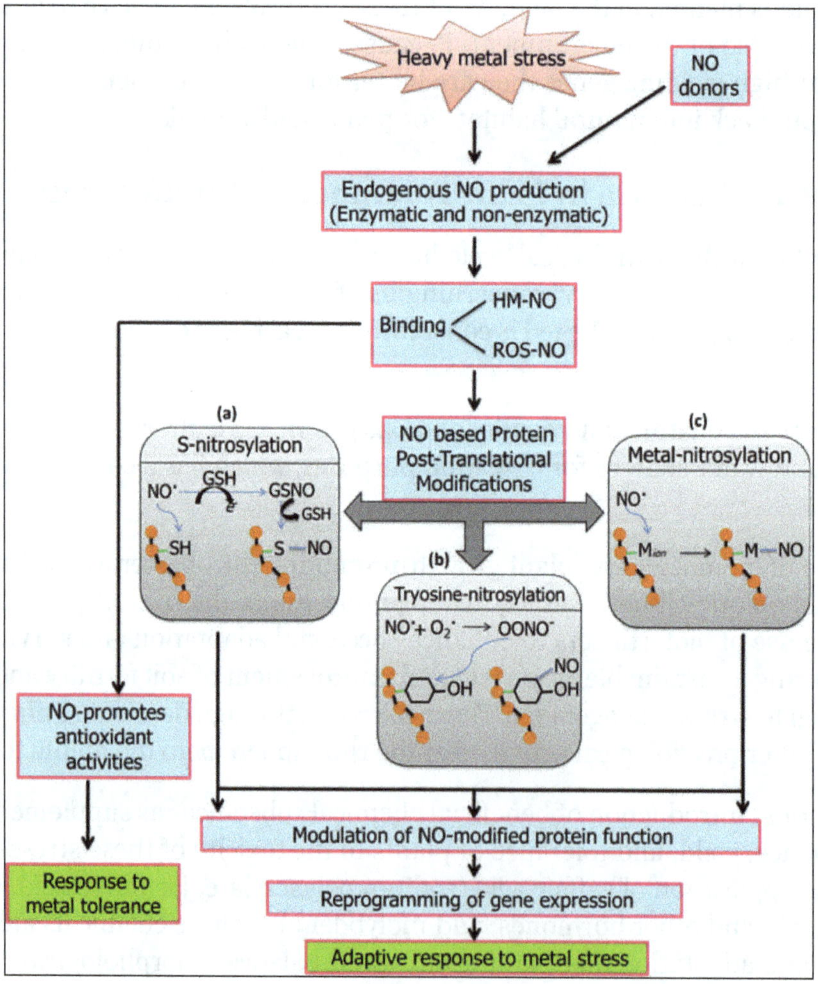

Sustainable Agriculture: The Influence of Climate Change and Human Population Growth on Crop Yield

The two major challenges affecting crop production in this modern time are climate change and escalation of human population number. The two are interrelated; one influences the other. As the human population increases, the need for fuel energy increases. Also, the need to generate heat for cooking food leads to deforestation. These human activities elevate the concentration of carbon dioxide in the atmosphere.

Climate change could result in an unregulated supply of rain water which could either occur in excess or in prolonged short supply. Whichever the case, crop production is affected negatively. Extreme rain fall gives rise to erosion and flooding, while drought results in loss and poor performance of crops. In a bid to boost crop productivity and yield and to combat drought-induced stress, farmers resort to the use of irrigation which adds salt to the soil. Drought is a great cause of poor yield of crops leading to food shortages.

Carbon dioxide increase, on the other hand, may aid the growth and nitrogen assimilation of plants. For example, although water limitation affects the clover plant's fixation of nitrogen, in the presence of a moderate supply of water, the plant could still perform excellently in the presence of carbon (IV) oxide and increased atmospheric heat.

However, adequate nutrient management with respect to improving the fertility of the soil could be an aid in solving the problem of climate change challenges on crops together with the use of rhizobacteria to boost crop production.

Population explosion also creates an urgent need for the establishment of shelter, recreation centers, industrialization, urbanization, etc. to accommodate human activities, and this affects land availability for mechanized farming.

To sustain the nutritional need of the growing human population, agricultural practices must be intensified, irrespective of short supply of rain water, environmental effects of the use of inorganic chemicals (fertilizers), and salinity issues. The biotechnological application of microbes may address the rising problem and ensure sustainability in the provision of food for all. In addition, the use of biofuels in automobiles as well as the practice of afforestation should be encouraged to help cut down on greenhouse gas accumulation for effective control of climate change. The use of electric automobiles should also be incentivized.

Plant Growth-promoting Rhizobacteria: The Redeemer of Plants

Plant growth-promoting rhizobacteria are those indispensable microbes possessing the unique abilities of supporting directly and indirectly the wellbeing of plants. These microbes, in order to survive in the rhizosphere, expanded their biological activities that influence the survival and growth of plants.

A number of these microbes with the enzyme machinery necessary for the breakdown of the exudates of plants are able to protect the plant from stress arising from water scarcity and salt pollution. They produce a variety of substances such as deaminase enzyme, plant hormone–indole acetic acid, PO_4^{2-} solubilizing enzyme, and microbiocidal/biostatic enzyme.

Some of these microbes contribute to plants nutritionally by trapping and integrating nitrogen into the plant via nitrogen fixation. However, as a means of surviving the stiff competition for existence and dominance at the root environment, some microbes devise a means of burrowing/penetrating the tissues of plants and establish themselves as endophytes within the plant, contributing not only to the nutrition of the plant but also to enhancing plants' survival rate and adaptation to their environment.

PGPR contribute to sustaining the intrinsic resistance of plant to pathogenic and environmental challenges. Some of these organisms are excellent in biofilm formation and secretion of polysaccharide substances which confer stability to plants during stress. As microbes contribute significantly in biohydrometallurgy by the generation of metal-rich solution through the biological oxidation of sulfur-containing ore, they also play a role in the immobilization of metals or cations and render them non-bioavailable using polymeric substances and other chemicals they produced (siderophores etc). Their presence can contribute to the reduction in metal stress on plants when applied to them as bioinoculants. The growth and survival of plants will not be possible without the help of these "farmers' friends" living both within and around the plant surfaces.

A healthy plant is a product of a healthy relationship between the plants and the growth-promoting microbes, while an unhealthy relationship could be observed in the degradation of the physical and physiological wellbeing of the plant. PGPR are the key players in the fight for sustainable plant development amidst stress conditions arising from climatic as well as manmade activities. The classes of microbes belonging to these group/genera are *Micrococcaceae HW-2* and many others.

Drought Stress in the Reduction of Plant Performance

Drought has been implicated in the reduction of plant metabolic and physiological activities. It reveals its effect in the degree of biomass yield, growth, nodulation, and chlorophyll greenness of plants. It is a major agricultural crop impairment inducer and a global concern as it affects the supply and sustainability of food availability to people and livestock. Rainfall is the major source of water for growing crops/plants by farmers (in a subsistent agriculture) in many parts of the world. The gradual delay or low level of rainfall due to the influence of climate adjustment to manmade activities is affecting the biotic aspect of the ecosystem particularly green plants which are the primary and major producers of food upon which all life forms depend for meeting their daily nutritional needs.

The effect of drought is obvious, yet the microbes domiciled at the root zone of plants, known as plant growth-promoting rhizobacteria, have proven to contribute to the tolerance and quick adaptation and adjustment of plants to drought stress via supplementary production of phytohormones that induce root growth, thus allowing a better water absorption, production of deaminase enzyme responsible for the disintegration of ACC molecules, nitrogen fixation, and solubilization of P.

Table: Contributions of microbes in the induction of drought tolerance by plants.

Plant species	Microbes	Microbial enhanced plant productivity under stress
Tomato	Rhizophagus irregularis Variovorax paradoxus 5C-2	Enhanced photosynthetic rate, reduced lipid oxidation, and increased root water conductivity and oxidative phosphorylation in the plant
Chickpea	Pseudomonas putida MTCC5279 (RA)	Reduced/controlled the expression of stress response gene, maintained water content, osmolyte, membrane structure, and germination rate of the plant

Maize	Azospirillum spp. (Az19)	Improve the growth and productivity of the plant under water stress compared to the control
Wheat	Piriformospora indica	Enhanced adaptation of plant by promoting nutrient and water absorption, improved root growth, biomass, water, chlorophyll and modulate the activities of antioxidant molecules
Trifolium repens	Rhizophagus intraradices Bacillus megaterium Pseudomonas putida	A consortium of these microbes increased plant nutrient and water contents, reduced stomatal conductance and stress enzyme activities for better adaptation to drought
Wheat	Azospirillum brasilense Sp245	Increased growth and expansion of xylem in the coleoptile of inoculated plant for easy conduction of water
Lavandula dentata	Arbuscular mycorrhizal fungi Bacillus thuringiensis	Co-inoculation enhanced plant growth, nutrient content, biomass, and reduction in lipid oxidation of the plant
Maize	Arbuscular mycorrhizal fungi Bacillus thuringiensis	Improved nutrient content and water transport protein as well as reduce lipid oxidation in the stressed plant
Brassica oxyrrhina	Pseudomonas libanensis TR1 Pseudomonas reactans Ph3R3	Enhanced growth, pigment and water content as well as phytoaccumulation of heavy metals in the plant
Maize	Pseudomonas putida (FBKV2)	Encouraged root and shoot growth, dried biomass weight and reduced stomatal conductance in the plant
Common Bean	Rhizobium	Promoted plant weight, nutrient content and increased Phaseolus vulgaris yield

The survival and enhanced influence of endophytic microbes toward poplar adaptation to water stress, for example, revealed that the presence of these organisms aided the physiological wellbeing of the plant and enabled it to tolerate water scarcity in the soil. The result showed 28% increase in biomass (shoot and roots) as against the control.

The increase in root number and tissue mass promoted the wider reach of plant roots for water and nutrient absorption, which enabled them to thrive in water scarce environments. This survival approach of plants is akin to the aid of microbial symbionts capable of plant hormone production (indole acetic acid, gibberellin), which stimulate growth and the resultant stress annulment for better performance of the plant.

In line with the survival-induced strategy, the endophytes will produce a poly-sugar substance known as trehalose, capable of protecting biologically produced compounds and molecules from breakdown during water induced osmotic tension.

It was observed that Ocimum basilicum L., the plant popularly known as basil, had improved chlorophyll pigment content and antioxidant activity under water-induced stress in the presence of mixed rhizobacteria consortium—Pseudomonas sp., Bacillus lentus, and Azospirillum brasilens (which indicated the aid of microbes in the synthesis of useful substances in the plant in spite of the prevailing stress challenges. Among the nitrogen fixers of atmospheric N to plants for its nutritional needs, Azospirillum is a good farmers' friend that contributes in enriching the soil and enabling plants to thrive under abiotic stress. This microbe, which possesses a list of unique attributes such as auxin production, deaminase enzyme production, siderophore–iron trapping substance, nitrogen fixation, exopolysaccharide, etc., enables its associates (plants) to tolerate drought as well as salt disturbances.

Azospirillum generates biocidal substances (bacteriocins, hydrogen cyanide, proteolytic enzymes, siderophores) capable of destroying pathogen/invaders of plants and encourages plants to tolerate abiotic stresses as well as living organisms inducing stress on plants. It could be regarded as a plant growth-promoting bacteria and is widely studied by researchers. The contribution of this organism in nutrient uptake by plant as well as water-facilitated uptake is well documented. Therefore, Azospirillum and other microbes capable of enriching and unbinding trapped nutrients and making them available for easy taking up by plants are called biofertilizer and they in turn depend on plant excretory products (root exudates) for their nutrition.

A closer look at the biosynthesis of siderophore by Gordonia rubripertincta CWB2 suggests that gorA gene under expression in Escherichia coli produce GorA hydroxylase enzyme (N-hydroxylating monooxygenase) which was able to hydroxylate the substrate putrescine (1, 4 diaminobutane) in the presence of coenzyme (FAD and NADPH). The hydroxamate is a crucial part in siderophore synthesis. It is responsible for binding to irons and other metallic elements in the rhizosphere.

A screening study on rhizobacteria obtained from cacti rhizosphere showed that the predominant isolate belonging to the *Bacillus* sp. exhibited amazing plant growth-promoting traits like the production of indole acetic acid, PO_4^{2-} solubilization, and hydrogen cyanide and NH_3 production. *Bacillus* has the attribute of producing exopolysaccharide substances that enable them to form biofilms and survive harsh environments, and has contributed to the survival, growth, and performance of *Zea mays* L. seedlings leaf area, stem, and biomass shoot dry weight in a simulated water stress experiment. This attribute of poly-sugar production by microbes strengthens the resistance of inoculated plants to water stress inhibitory effects and thus ensures their growth and survival in a water-deficient environment.

To understand the consequences of water deficiency on plant physiological condition and activities, groundnut plants were subjected to water stress using a hydrocarbon compound—polyethylene glycol, which neither penetrated nor posed any toxicity effect to the plant, rather it caused a decrease in the plant RNA, chlorophyll composition, and water content shortfall in the plant. The observed effect is concentration based. The higher the concentration of the polyethylene glycol, the higher the water stress effect it exerts on the plant and the higher the corresponding reduction effect of RNA in the leaves and roots. This effect was observed in the quantity of water content of the plant leaf (relative water content). The drought interfered with chlorophyll *a* more than *b*. The production of oxygen radicals and hydrogen peroxide molecules are responsible for destruction of lipids and the breakdown of chlorophyll which leads to reduction and yellowing of the leaves.

In another experiment, a legume plant (Aspalathus linearis) was subjected to drought condition via withholding of water supply to the plant. This led to an observable effect on the photosynthetic rate of the plant by 40% reduction as well as 61% reduction in stomatal conductance, as a consequences of continuous closure of stomata to maintain intracellular water content and cut down on water loss from the leaves. This is one of the strategies plants adopt to increase their efficiency in the use of water during drought stress periods. Drought-induced closure of stomata has a direct connection with the reduction in plant assimilation and fixation of carbon (IV) oxide from the atmosphere. Inadequate supply of water directly impedes the process of photosynthesis in plants. Preferential development and growth of underground plant parts (roots) to shoot parts of the plant undergoing stress was also observed, which enabled the plant to absorb more water from the inner layer of the earth.

Although A. linearis is capable of forming a symbiotic relationship with rhizobium microbe, drought, on the other hand, affected the nitrogen nutritional content of the plant as a result of its influence on the activities of the nodules and the N_2 fixing enzyme (nitrogenase). This consequently makes the plant more dependent on the available nitrogen in the soil.

Periodic or interval water supply through rain water to rice farming has proven to contribute to water stress in rice plants. Rice, being a moderately water-loving plant, finds it difficult to cope with dryness arising from drought and requires a suitable measure for increasing the adaptability and productivity of the plant in a low water vicinity. The application of potassium salts as nutrient (potassium chloride and potassium sulfate) could be a good measure in reducing water stress on rice plants. Potassium nutrient applied at a concentration of 120 kg per hectare was able to increase rice yield and the index of its harvest within 15 days of water scarcity. Potassium has a link with the efficiency of water use by the plant and improves it via the promotion of transpiration-pull-boost with the assistance of potassium ions which are components of cell membranes mediating active transport of nutrients as well as water into the rice plant.

The ease of water entry from the soil into the plant contributes in canceling the potential damage of tissues by oxygen radicals during water-deficient condition. A 338% increase in proline concentration within the plant treated with K^+ salt was observed.

The use of sugar beet fermented by Aspergillus niger as soil amendment in the presence of bioinoculant Bacillus megaterium and a consortium of arbuscular mycorrhizal fungi promoted plant growth, biomass, nutritional content, and water level and reduced the conductance of stomata and the activities of enzymatic antioxidants in plants faced with drought stress. The amendment boosted the effectiveness of the bacteria toward supporting the growth and productivity of the stressed plant.

The importance of water as a medium for transport and biochemical reaction in a living system cannot be overemphasized. A limited water environment interferes with the smooth biochemical processes and interaction of molecules in the plant as well as its associated microbes in the rhizosphere. The performance of Glycine max under drought stress was increased in the presence of Bradyrhizobium japonicum and a nodulation stimulant—genistein. Genistein as well as daidzein iso-flavone substances are capable of inducing response reaction of nodule gene expression in Bradyrhizobium sp. These compounds are components of exudates from legume roots. Genistein treatment of G. max seed together with a bioinoculant (B. japonicum) greatly encouraged the resistance of soybean to drought stress and decreased its detrimental effect on the formation of nodules by the microbe.

The tolerance of plants to drought stress is dependent on their own genetic makeup. The physical manifestation observed in a plant is in direct correlation with the quality and quantity of the physiological activities taking place at the gene level. All biomolecules produced are the product of gene expression. The easier the expression of genes encoding the biomolecules that protect the plant cellular component from drought induced cellular destruction/damage, the more the viability and productivity of the plant at a given stress condition. Different species of soybean undergoing a 15-day vegetative growth under drought stress revealed that the electrolyte loss was more in plant tissue stressed for 10 days and above. The shoot dry matter or biomass concentration decreased (by 79.18%) at the 10th day to 15th day of stress, while the root biomass or dry weight/root growth increased (by 100.96%). This means that drought promotes the accumulation of nutrient and photosynthetic products at the root and stimulates its growth and development to maintain balance, adjustment, and sustainability of plant physiological adaptation to the stress problem. Soybean species (SJ-4) possessing unique genetic/physiological characteristics performed better than others under drought stress and should be the crop of choice for drought-predominant locations.

Also, two soybean species showed slight yield reduction of 8 and 12%, respectively, as compared to when grown under consistent water supply via irrigation. This implies that they are good in tolerating drought as a result of their inherent delay in wilting and a corresponding elongation of fixation of carbon dioxide in their leaves.

In the same vein, the accumulation of amino acid (proline) in the leaves of tomato plant and increased value of the relative water content of the leaves of the plant confer resistance to drought by tomato plant. Proline is responsible for sustaining the movement of water molecules from a region of higher concentration to that of a lower concentration in response to concentration gradient between the plant and its environment. It helps the plant maintain its turgor intracellular pressure by facilitating movement of available water from the soil into the root of the plant.

A non-bioinoculant phyto-drought-stress-mediated strategy involving the application of chitosan to a plant (thyme) undergoing drought stress helped to offset the effect of drought on the plant by 20% compared to the untreated control which showed a decrease in dry matter of the plant by 54 and 56% on the photosynthesis retardation of the plant under drought stress. In the same manner, it encouraged the synthesis of essential oil in the thyme plant, reduced the degree of peroxidation, sustained the cell membrane conformation and function, and stimulated the accumulation of proline amino acid in the leaves of the plant by 20%. Being a polysaccharide, chitosan is generated from chitin alkalinization of N-deacetylation. These molecules are abundant in the outer skeletons of insects, fungi, and algae cell wall composition and are implicated in secondary metabolite production in plants when applied to them.

A sainfoin plant inoculated with arbuscular mycorrhizal fungi exhibited 7.27, 4.21, and 2.40% increase for relative water content, N_2, and P leaves total content, respectively, at the 40th day of the experiment. This fungus enabled the plant to adapt and survive during drought limitations by enhancing plant growth and protecting it from damage induced by stress. Generally, an external–internal symbiotic relationship by soil fungi (arbuscular mycorrhizal fungi) performs the same protective role as endophyte does to plants. There is a biomass increase as well as a decrease of Fusarium oxysporum infectivity of tomato plant inoculated with Glomus intraradices and Piriformospora indica irrespective of the nutritional content of the soil used. Also, the involvement of mycorrhizal fungi in nutrient dissolution/solubilization particularly as it concerns phosphorus gives credence to its participation in the direct promotion of plant health and productivity.

These microbes form a channel (protein transport channel) for dissolution and absorption of solubilized PO_4^{2-} from the rhizosphere into the host plant in a mutualistic dependent relationship. They are generally used as biofertilizers for immobilized nutrient dissolution via the organic acids produced which are capable of reacting with the compounds of trapped minerals and converting them to soluble forms for easy assimilation by plants.

The presence of mycorrhizal fungi adds further surface area for water and nutrient acquisition, thereby making the plant more resilient and tolerant to the climate change-induced stress. This notwithstanding, they also partake in structure building, arrangement, and improvement of soil for proper aeration and migration of water together with nutrients in the soil, making the soil healthy and fertile. It is a known fact that fungi can survive and thrive well in a dry or semi-dry soil and still perform their normal activities. This property is exploited by arbuscular mycorrhizal fungi in making an inoculated plant perform its activities under drought condition as the organism reaches out for more water within the soil where the roots of the plant can possibly not reach.

In addition to fungi solubilizing PO_4^{2-} bacteria, the enzymes (phytase, phosphatase) can produce organic acid (acetic acid, citric acid) and will be able to mineralize or solubilize P via ionic interaction of the charged group in these molecules. These solubilizers of P play a part in the enrichment of soil fertility and supporting plant nutritional requirements irrespective of the prevailing bioavailability or non-bioavailability of the nutrient in the soil. Though the level of available P can determine the function of the microbes as it relates to solubilization of P, the higher the available nutrients, the less the quest to solubilize already immobilized phosphorus and vice versa. Microbes are the key players in the replenishment of P in the soil.

Also, bacterial inoculation of rice plant using *Pseudomonas fluorescens* promoted the intrinsic tolerance of rice plant to drought stress and encouraged the expression of abscisic acid synthetic genes particularly at the stage of reproduction by the plant. This confers an induced systemic plant resistance to drought stress. It implies that microbes, especially the PGPR colonizing the rhizosphere of plants, have an indirect as well as a direct role in enhancing the expression of genes by water-deficient plants through a process known as induced systemic resistance. Microbes in their various capacities have been shown to aid plants in their tolerance to drought.

The symbiotic N_2 fixers (*Sinorhizobium meliloti* and *Sinorhizobium medicae*), on the other hand, possess the ability of protecting a legume (*Medicago truncatula*) from senescence of the leaves under drought stress. The rhizobial inoculants aid in the delay of the process of senescence by promoting the accumulation of potassium ions, reduction in the protein mediating ethylene production, and induction of cytokinin production which inhibit senescence and accumulation of sugars and amino acids that aided in the plant survival during stress.

Nodulated medicago plants recovered from drought stress faster than the non-nodulated legumes as a result of the shifts in the carbon partitioning from starch to sugars and thus the enhanced allocation of reserves to osmolytes during drought, enabling to stay green with the ability of fast recovery after rewatering.

Drought stress reduced relative water content and chlorophyll and increased proline concentration of the plant. Obviously, drought affects root, shoot, yield, and overall performance of the affected plant. Yet at moderate water scarcity or even severe conditions, the intrinsic quality of the plant determines to a large extent its sensitivity or tolerance to the drought condition. The osmotic adjustment and performance of three barley plant species was measured when they were subjected to different drought treatments (moderate and severe). The results showed that yousof species possesses a drought tolerance trait and was able to have an increased root length to shoot length ratio under severe stress conditions.

The test plants survived the water stress by employing an adjustment in the osmotic behavior of their root systems leading to the accumulation of solutes such as proline in their cells to maintain the cell structure and function during water scarcity. This adjustment in osmosis to sustain the turgidity of the plant cells directly influences the escalation level of photosynthesis and tissue growth of plant during drought stress.

The symbiont (Bradyrhizobium sp.) aided the tolerance of cowpea plant to water scarcity stress and boosted the quality of NO_3^- (nitrate) and amino acid (proline) in the inoculated plant.

Groundnut inoculated by Bradyrhizobium under drought conditions had a large quantity of amino acids which was derived from the nitrogenase-catalyzed conversion of atmospheric N to NH_4^+ ions necessary for amino acid and protein formation in the plant. The observed increase in protein level of legume-inoculated plants implied that these microbes aid in the supply of nitrogen via fixation in the presence of the nitrogenase enzyme operating at the root nodules of the plant. Bioinoculation of legume plant with rhizobium has helped to encourage the development of many leaves in the plant due to the greater number of root nodules in the plant roots that aid in nitrogen fixation.

The treatment of annual medic plant growing in sufficient water (irrigated) and water-deficient (dry farming) systems with a mixture of biofertilizer and chemical fertilizer in an integrated fertilization practice encouraged the plant to efficiently adapt to the water-deficient condition and accumulate both macro- and micronutrients in the tissues of the plant compared to the use of chemical fertilizer alone. The application of bioinoculants and mineral fertilizer to a plant growing in a water-limiting environment is more effective in boosting plant productivity than in water-sufficient soil.

Table: Plant drought stress tolerance mediated by synergy between microbes and soil amendment.

Plant species	Microbes	Amendment	Plant productivity and tolerance outcome
Arundo donax	Micrococcaceae HW-2	Sodium polyacrylate	Microbial attributes of plant hormone, deaminase and siderophore production and enhanced water retention capacity of sodium polyacrylate promoted shoot growth, biomass, and root of the plant
Thymus vulgaris Santolina chamaecyparissus Lavandula dentate Salvia officinalis	Enterobacter sp. Bacillus thuringiensis Bacillus megaterium Bacillus sp.	Fermented agrowaste	The amendment enhanced nutrient uptake via bacterial stimulated activities for proper nutrient absorption by plants and stomatal conductance during drought stress
Pinus halepensis	Azospirillum brasilense Pantoea dispersa	Olive-mill waste	The treatment and the inoculants promoted the carbohydrate and microbial biomass carbon as well as soil nutrient and consequently increased growth, water content and nutrient uptake by the plant
Lavandula dentata L.	Bacillus megaterium Enterobacter sp. Bacillus thuringiensis Bacillus sp.	Composted sugar beet	It increased biomass shoot dry weight, root, and nutrient content of the plant. The amendment increased the concentration of bioavailable phosphorus and nitrogen in the plant rhizosphere
Cistus albidus L.	Azospirillum brasilense Pantoea dispersa	Olive residue	Increased dry root, shoot weight, organic carbon, soil enzymes, and microbial biomass carbon

The attenuation capacity of microbes to the negative effects of drought on plant development was facilitated by the use of a combination of Azotobacter chroococcum and Pseudomonas fluorescence together with phosphorus fertilizer, and was able to boost the inorganic P content of the plant (soybean) growing under insufficient and abundant water content of the soil. The N content of the leaves and root of the soybean plant were increased by 6 and 8% under water stress condition. The absorption of phosphorus in the soybean plant under the influence of bioinoculant increased by 16%.

Although drought negatively affected plant development such as water content, chlorophyll of the leaves, and the dry weight of the root with a corresponding increase in the amino acid proline and abscisic acid–osmoprotecting molecules, yet the application of humic acid to the drought-stressed plant has reversed the observed effect by decreasing the quantity of proline and absicsic acid content of the treated plant, with resultant increases in chlorophyll content, dry weight of the root, and water content of the pistachio plant during drought stress circumstances.

However, it is clear that bioinoculants and the application of useful chemicals and amino acid could help plants avert the challenges of drought conditions and aid in the promotion of a sustainable production of crops especially in those regions prone to water scarcity (insufficient rain fall).

Salt-induced Challenges on Plant

The deleterious nature of sodium chloride and other salt compounds on the growth and development of plants particularly in inducing water limitation stress and uncontrollable negative stomata closure cannot be overemphasized. This necessitates the application of an attenuation strategy to cancel the effect of salt on crops.

The co-inoculation of Pseudomonas and endomycorrhizae on cowpea plant undergoing salt-induced stress showed a decrease in the mycorrhizal infection of the plant. Treatment of 6000 ppm sodium chloride in the presence of the fungi and bacteria increased the carotenoid concentration (0.449 mg/g). Also, osmolytes (proline and sugars) increased in the presence of endomycorrhizae inoculation in the plant. Irrigation with tap water in the midst of endomycorrhizae and Pseudomonas fluorescens gave higher fresh and dry weight, pod length, seed number, and protein content of the cowpea plant compared to the one irrigated with salt water.

As part of microbial facilitated plant tolerance to salinity, biofilm formation is a suitable strategy which microbes employ in enhancing barley plant tolerance to soil salinity. This was observed in the aversion/reversal of the harmful effect of salt on a number of growth parameters such as seedling length, relative water content of the plant's leaf as well as fresh and dry weight of barley plant. This biofilm formation strategy by microbes is actually a means of protecting themselves from the harsh conditions they found themselves in (short supply of nutrient, ionic toxicity as well as water limitation induced osmotic stress), enhancing survival within the presence of these limited resources by aggregating in masses (sessile compartmentalization) of cells at both living and non-living surfaces present at the root environment and contributing indirectly to tolerance ability of the plant to stress.

Bacillus amyloliquefaciens (SQR9) contributed to maize salt tolerance by enhancing its chlorophyll production level via colonizing and interacting with the plant roots. It also aids in the exclusion of sodium from the plant root, and stimulates the production of sugars and antioxidant within the tissues of the maize plant. This implies that a sensitive plant could become tolerant to salt stress in the presence of a bioinoculant.

This unique contribution to the wellbeing of the plant could be ascribed to its ability to produce indole acetic acid, solubilize phosphorus, colonize the plant root, and produce enzyme deaminase (ACC-deaminase) capable of maintaining optimum biological function of the plant in the midst of stress inducers.

Table: The unique contributions of microbes to salinity tolerance by plants.

Plants	Microbes	Microbial influenced plant productivity and salinity tolerance
Chili	Bacillus spp. Alcaligenes spp. Proteus spp. Aneurinibacillus aneurinilyticus	Significantly increased root and shoot length more than the control
Rice	Enterobacter sp.	Promoted the growth of rice seedling and reduced ethylene production and antioxidant enzyme activities in the plant
Rice	Bacillus sp.	Aided the alleviation of salt stress by increasing the biomass and growth of rice seedling via production of indole acetic acid and deaminase enzyme
Festuca arundinacea	Enterobacter ludwigii	Membrane transport protein in the microbe that control sodium and hydrogen ion movement across bacteria cell and the production of plant hormone, phosphate solubilization, nitrogen fixation contribute towards the growth, tolerance, and plant productivity
Alfalfa	Halomonas maura Ensifer meliloti	Increased the weight of shoot dry weight, yield, and water content of the plant
Rice	Thalassobacillus denorans (NCCP-58) Oceanobacillus kapialis (NCCP-76)	Inoculated plant was observed to have increased germination ability, root and shoot growth, protein, and chlorophyll contents as well as nutrient contents with reduced sodium ion accumulation in the plant
Rice	Bacillus pumilus	Enhanced plant growth and decreased the accumulation of sodium ions without having an effect on boron accumulation in the leaf tissues
Oat	Klebsiella sp.	It boosted plant growth, water content, dry shoot, and root weight of inoculated plant
Wheat	Bacillus subtilis	Enhanced salicyclic acid content of the plant, leaf water content, and reduced proline and malondialdehyde content of the plant for better induction of systemic resistance
Lettuce Radish Chinese cabbage	Arthrobacter scleromae SYE-3	Increased all the plant shoot length and leaf number in lettuce by 45.1%

With a shift in perspective, it was understood that opportunistic pathogens of humans can aid in plant tolerance to salt stress. The organism Methylobacterium mesophilicum could efficiently colonize and dominate the root rhizosphere region of three plants (cucumber, tomato, and paprika) under the study at salt concentration as high as 6% and possess the capability of phytohormone (IAA) production together with fungicidal enzyme production that contributes to the growth, protection, and tolerance of the plant to biotic and abiotic stress.

Sensitive crops have remarkable reduction in viability and performance when challenged with salt stress as compared to tolerant species, following the observed accumulation of proline and reduced leaf area in a sensitive wheat plant compared to a tolerant plant. However, the application of synthesized phyto-hormone (gibberellic acid) on salt-stressed wheat plants has resulted in adequate adaptation of two wheat plant species to salt-influenced stress and improved its yield.

Moreover, the efficiency of microbial exopolysaccharide formation and the formation of biofilm during environmental stress at the plant rhizosphere is a contributory factor toward plant adjustment to salinity or salt-induced osmotic shock. The polysaccharide molecules produced by microbes in a bid to protect themselves from desiccation and ionic toxicity have a charged part which reacts with dissociated sodium ions and chelates them. This reduces the toxicity and abundance of these ions in the rhizosphere, making the soil suitable for root proliferation.

The quest for a better way to reduce or alleviate stress inducers has led researchers to investigate the use of non-biological methods such as the external application of micronutrient Mo (molybdenum) as an additive to plants facing salinity stress. When molybdenum was applied to a bean plant (Phaseolus vulgaris) subjected to salt stress, the plant was able to maintain its chlorophyll composition and tissue components. Mo being a unique player (cofactor) in cellular enzyme biosynthesis and activities enhanced the formation of chlorophyll and maintenance of adequate cellular function.

The critical evaluation of the microbial community of Aster tripolium L. showed the predominance of gram-positive bacteria residing in the root of the halophyte, while gram-negative bacteria exist more in the rhizosphere and bulk part of the soil. The isolated organism exhibited the ability of forming 1-aminocyclopropane-1-carboxylate deaminase, indole acetic acid, etc. that contribute to the survival and tolerance of the plant to salt-induced stress and toxicity.

It was noted that rhizosphere bacteria of a desert-adapted plant (Suaeda fruticosa) from Kutch Desert has a unique strain of bacteria (Bacillus licheniformis strain A2), which solubilized phosphate, produced indole acetic acid, siderophore etc., and contributed to 31% groundnut height and 43% biomass increment of the plant and 24 and 28% rise when grown in 50 mM sodium chloride supplemented soil.

Another non-biological stress alleviation technique showed that caffeic acid was able to aid the plant (soybean) tolerate salt stress. The applied caffeic acid stimulated the NO (nitric oxide) composition of the nodules, which has a direct relationship with the induction of cyclic guanine monophosphate responsible for controlling reactive oxygen radicals produced in the stressed plant. Caffeic acid helps to shield nitrogenase enzyme and leghemoglobin of the soybean root nodules from the destructive effect of sodium and chloride ions, thereby adding to the growth of the plant by enabling the symbionts to fix nitrogen adequately to meet plant nitrogen requirements. Caffeic acid has also been implicated in chlorosis suppression of treated plants undergoing salinity-induced stress.

On the other hand, it was demonstrated that sodium chloride impedes the uptake of potassium ions by Broussonetia papyrifera plant, resulting in high sodium to potassium ratio. Sodium ion equally decreased the calcium, magnesium, and phosphorus content of the plant root system and caused an induced ion imbalance in the plant rhizosphere. Bio-active enzymes responsible for controlling oxygen radicals (catalase, peroxidase, superoxide dismutase) were also decreased by the inhibitory effect of the high salt concentration in the plant environment. Protein content of the plant was also affected as observed in the appearance and disappearance of protein bands in the plant samples analyzed. Plant growth-promoting rhizobacteria inoculated on wheat plant promoted its root and shoot growth and the weight of the fresh tissue component of the plant. The observed rates of plant growth were from 62.2 to 78.1% in the presence of the bioinoculant. This could be attributed to the production of phyto-hormone by the microbes. Stress-adapted bacteria (Pseudomonas putida R4 and Pseudomonas chlororaphis R5) capable of producing plant hormone (IAA) at the rhizosphere of inoculated cotton plant subjected to salt stress completely averted salinity stress on cotton plant. This was also found in cotton plant inoculated with Klebsiella oxytoca (Rs-5) that resulted in the increase in seed germination (15.40%), growth, and overall tolerance of cotton to salinity stress.

A newly identified rhizobacteria (Ochrobactrum intermedium) possessing the survival ability of changing its membrane phospholipid composition when subjected to stress condition and also producing indole acetic acid, siderophores, deaminase enzyme, and utilization of nitrate happens to improve the growth (root and shoot) of groundnut plant subjected to salt stress better than a well-known classified Bradyrhizobium sp. accepted as a suitable groundnut inoculant by the Argentine INTA organization. The organism also performed better in the production of biofilm than Bradyrhizobium and recorded high tolerance to 300 mmol sodium chloride and higher temperature.

Soil additives containing a variety of organic matter (poultry droppings, olive plant waste, rice straw, molasses) and a consortium of microbes known as "effective microbe" (bacteria, yeast, photosynthetic bacteria, actinomycetes, etc.), collectively regarded as bokashi, have proven to be effective in shielding Mandarin tree (citrus plant) from salt stress and improving its productivity, and the nutritional content of the fruit, and also in enhancing the rhizosphere microbial community count and population. The bokashi treatment has aided greatly in the salted soil fertility improvement and protection of the plant from osmotic salt-induced stress.

In a salt toxicity alleviation study, microbial species known as Kocuria erythromyxa EY43 and Staphylococcus kloosii EY37 were found to be very potent in increasing strawberry plant growth, chlorophyll, mineral content, and fruit yield. The organism was able to inhibit the absorption of toxic ions (sodium and chloride ions) from the plant rhizosphere making a salinity-sensitive strawberry plant insensitive and tolerant to the condition of salt stress. The integration of endophytic bacteria (Sphingomonas sp. LK11) with applied jasmonic acid synergistically improved the overall tolerance of tomato plants (wild type and mutant species) to salinity. The combined applied microbe–jasmonic acid enhanced root and shoot growth of the plant, brought about proper regulation of glutathione content of the plant, and enabled it overcome the detrimental effects of salinity. It was noted that the microbial-jasmonate treatment lowered the intracellular abscisic acid level of the plants.

Although application of nutrients or minerals is an alternative measure to curbing salinity stress, introduction of excess of these nutrients will result in phytotoxicity instead of phytoenhancement and will necessitate remediation.

Three hundred percent growth in barley plant growing in salt-containing soil was recorded when inoculated with Curtobacterium flaccumfaciens which did not fully express the attributes of plant growth promotion in pure culture growth medium. This organism, being relatively new in this branch of study, was originally associated with the infection and destruction of crops but was now found to protect plants (barley) from salt stress better than the well-known microbes exhibiting excellent characteristics of plant growth promotion such as Microbacterium natoriense and Pseudomonas brassicacearum.

Rhizobium naturally adapted to desert environments has been shown to perform well in establishing nodules on the root of legumes at a high level of salt–soil content and to contribute to growth and better performance of the legume. Also, an endophytic bacteria (*Bacillus subtilis* LK14) isolated from the plant (*Moringa peregrina*) bark possesses high indole acetic acid production capacity and ACC-deaminase was able to improve the growth and development of tomato plant seedlings.

Microbial facilitated survival of plants under saline tension is dependent on microbial production of deaminase enzyme (1-aminocyclopropane-1-carboxylate deaminase) responsible for degrading the plant product known as aminocyclopropane-1-carboxylate, the precursor of the plant hormone (ethylene). This hormone is responsible for the interference of root development when challenged with stress substances that induce elevated level of ethylene production and is well documented by Tak et al.

Chickpea plant affected by an increase in salinity was rescued by co-inoculation with two deaminase enzyme-producing microbes (Mesorhizobium MBD26 and Rhizobacteria RHD 18) with an observed nodulation capacity of 49 nodules per plant, 201 mg weight of nodules, 12.28 mg per plant nitrogen, and a rise in 31.2% of the above the soil plant part dry weight (shoot dry weight). The result was further increased to 53 nodules and 116.9% grain produced, with N_2 level spanning between 9.59 and 27.36 mg per plant investigated.

The production of deaminase by rhizobacteria, on the other hand, encouraged the root growth of velvet beans undergoing drought stress by catalyzing the disintegration of 1-aminocyclopropane-1-carboxylate to generate nutrients for the microbes and in turn create a shortage of the ingredient (ACC) necessary for the production of ethylene. This helped to limit the action of this hormone. The inoculation of *Bradyrhizobium japonicum* together with *Bacillus amyloliquefaciens* enabled better nodule formation on the soybean roots by the *B. japonicum* and resulted in better fixation of nitrogen and plant viability.

Alkalinity Stress in Phyto-retardation

Apart from stress induced by sodium chloride, sodium carbonate (Na_2CO_3) and sodium hydrogen carbonate ($NaHCO_3$) also constitute a problem to crops. They are implicated in the formation of alkaline soil that results in elevated soil pH and interferes with the bioavailability of phosphorus, iron, copper, manganese, and zinc resulting in induced nutrient deficiency and osmotic stress capable of interfering with the proper biological function of the plant. The high pH has its own inhibitory challenges on non-alkaliphiles inhabiting the rhizosphere. It also interferes with their biological activities as well as physiological functions of the plant.

However, fertility issues of alkaline soil could be handled by the application of bioinoculants. These microbes ameliorate the alkalinity effect on plant by supporting increase in the number of nodules formed. They boost nitrogenase enzyme activities for efficient nitrogen fixation and improve mycorrhizal dominance in the root of faba bean inoculated with *Rhizobium leguminosarum* and mycorrhizal fungi. The team work between the two organisms promoted faba bean productivity and resistance to alkalinity stress.

INTERACTION OF PLANT BIOTIC AND ABIOTIC STRESSES

Plants have evolved to live in environments where they are often exposed to different stress factors in combination. Being sessile, they have developed specific mechanisms that allow them to detect precise environmental changes and respond to complex stress conditions, minimizing damage while conserving valuable resources for growth and reproduction. Plants activate a specific and unique stress response when subjected to a combination of multiple stresses. In light of this, current techniques for developing and testing stress-tolerant plants by imposing each stress individually may be inadequate. This is particularly true for combinations of biotic and abiotic stresses, the signalling pathways of which can act antagonistically. While valuable, the results of studies investigating stress factors in isolation do not explain the effects of more than one stress on plants. There is an urgent need for a change of focus in plant stress research, in order to understand the nature of multiple stress responses and to create avenues for developing plants that are resistant to multiple stresses yet maintain high yields.

Challenge of Multiple Environmental Stresses in Agriculture

Most crop plants grow in environments that are suboptimal, which prevents the plants from attaining their full genetic potential for growth and reproduction. This is highlighted by analysing the difference between maximum crop yields and the average yield for that crop. For example, US wheat yields in a record year can be up to eight times as great as the average yield. The yield difference can largely be explained by unfavourable environmental conditions, which, when creating potentially damaging physiological changes within plants, are known as stresses. Abiotic stress factors such as heat, cold, drought, salinity, and nutrient stress have a huge impact on world agriculture, and it has been suggested that they reduce average yields by >50% for most major crop plants. Further to this, plants must defend themselves from attack by a vast range of pests and pathogens, including fungi, bacteria, viruses, nematodes, and herbivorous insects. Each stress elicits a complex cellular and molecular response system implemented by the plant in order to prevent damage and ensure survival, but often at the detriment of growth and yield.

Current climate prediction models indicate that average surface temperatures will rise by 3–5 °C in the next 50–100 years, drastically affecting global agricultural systems. This will be concurrent with an increased frequency of drought, flood, and heat waves. In particular, warmer, drier summers in mid-continental regions such as central Europe and central Africa are predicted, along with a reduction in growing season in many regions, extensive salinization as sea levels rise, and a decrease in land suitable for agriculture. A change in variability of rainfall and temperature may itself affect yields as well as adversely affecting nutritional quality of crops. Climate change will also influence the habitat range of pests and pathogens, with increasing temperature facilitating pathogen spread. Crop plants are therefore likely to encounter a greater range and number of environmental stresses, which when occurring simultaneously can have severe consequences. The changing climatic conditions, combined with an increasing pressure on global food productivity due to population increase, result in a demand for stress-tolerant crop varieties that has never been greater. Understanding the mechanisms of plant responses to multiple simultaneous stresses is therefore crucial in providing opportunities for the development of broad-spectrum stress-tolerant crops.

Importance of Studying Plant Stresses in Combination

The ultimate goal of creating stress-tolerant crops either transgenically or through conventional breeding has pervaded almost all aspects of plant science, and is pursued by both public and private sector researchers. Success has already been achieved in many areas. For example, resistance to abiotic stresses has been demonstrated in a variety of crops through genetic manipulation of transcription factors (TFs), late embryogenesis abundant (LEA) proteins, and antioxidant protein. Resistance to plant-parasitic nematodes has been conferred to potato, rice, and banana plants by the expression of transgenes. In addition, tolerance to both biotic and abiotic stresses has been achieved. In maize, breeding programmes have led to plants which are tolerant to drought and have additional resistance to the parasitic weed Striga hermonthica. However, research programmes aimed at developing tolerance to a particular stress do not necessarily test susceptibility to other biotic or abiotic stresses. This can have unforeseen consequences, as improved varieties may respond unpredictably when grown in field conditions. Examples include: a variety of cassava bred for resistance to Cassava mosaic virus (CMV), which was later found to be a greater target for the whitefly pest Bemisia tabaci; new drought-tolerant aerobic rice varieties that were particularly susceptible to nematode infection, perhaps due to their increased root length; and transgenic cotton plants expressing the Bacillus thuringiensis (Bt) insecticidal protein Cry that showed a reduction in the level of the protein during periods of high temperature, elevated CO_2 levels, or drought, leading to decreased resistance to pests. Transgenic plants overexpressing a proline biosynthesis enzyme (P5C) accumulate this osmoprotectant during drought and are thus resistant to osmotic stress. However, Rizhsky et al. found that under a combination of drought and heat stress, plant cells accumulated sucrose instead of proline, perhaps to protect hyperactive and therefore susceptible mitochondria from the build-up of potentially toxic P5C. In the transgenic plants an added heat stress may therefore counteract any osmoprotective benefit.

An increasing volume of evidence from field, laboratory, and molecular studies suggests that plants respond to a specific combination of stresses in a non-additive manner, producing effects that could not have been predicted from the study of either stress individually. For example, a report by Mittler *et al.* described that in the US between 1980 and 2004 the total agricultural losses attributed to drought were worth US $20 billion, but total losses due to drought combined with a heat wave totalled US $120 billion, suggesting that the presence of a second stress factor can exacerbate the detrimental effects of the first. Therefore, in order to develop crops that thrive and maintain a high yield in field conditions, an integrated approach should be adopted whereby resistance traits are tested under a range of stress treatments. It has been proposed that truly to characterize the response of plants to multiple stresses, it is crucial to impose the stresses simultaneously and treat each set of environmental conditions as an entirely new stress.

Plants must produce a tailored response to specific multiple stress conditions, as in many cases the individual stresses would normally elicit opposing reactions. For example, heat stress usually causes plants to open their stomata in order to cool the leaves, but under drought conditions this would be disadvantageous as more water would be lost. Similarly, increased transpiration caused by heat stress could enhance uptake of salt or heavy metals, heightening the damage from these factors. Plants must efficiently balance resource allocation between growth and defence against stress, as responding to stress can be costly and reduce fitness in terms of growth and yield. This cost is highlighted by examples where the experimental transgenic expression of a stress-responsive gene confers stress tolerance but to the detriment of growth and yield. For instance, constitutive overexpression of Arabidopsis thaliana DREB1A or rice DREB1A confers freezing and dehydration tolerance to Arabidopsis, but results in severe growth retardation. The constitutive expression of a pigeonpea proline-rich protein in Arabidopsis confers tolerance to osmotic, salt, and heat stress, but produces plants that are stunted in size. The cost to plants of defence is therefore likely to be reduced if specific genes and compounds have roles in several different stress responses, a potential explanation for the overlap between stress response pathways. This would explain the findings that certain molecular signalling pathways, TFs, and effector proteins are activated by both biotic and abiotic stresses, and that secondary metabolites such as flavonoids accumulate in response to and provide protection from a wide range of stresses.

The occurrence of simultaneous biotic and abiotic stresses presents an added degree of complexity, as the responses to these are largely controlled by different hormone signalling pathways that may interact and inhibit one another. Often, the exposure of plants to a pest or pathogen increases the effects of an abiotic stress such as water deficit, whereas long-term abiotic stress can weaken plant defences and cause enhanced pathogen susceptibility. Despite this, few reports in the literature have focused on understanding the interaction between biotic and abiotic stresses.

Transcriptome Analysis of Multiple Stress Responses

The investigation of plant molecular responses to multiple stresses has often focused on overlapping transcriptional patterns. To this effect, several studies have been carried out in which different groups of plants are exposed to either one stress or another in parallel, and their gene expression patterns compared. Overlapping sets of genes that are regulated by both stresses are then identified and proposed to represent a generalized stress response or points of cross-talk between signalling pathways. It has been speculated that these genes may be targets for improving stress tolerance in crop plants. For example, following comparison of the transcriptome response of *Arabidopsis* plants to nine different abiotic stresses using Affymetrix ATH1 microarrays, 67 genes were found to be commonly regulated by every stress, suggesting that there was a universal component of the response to each condition. A similar experiment investigated the expression of cytochrome P450 genes in response to various abiotic stresses, pathogens or hormone treatments. Genes were identified that were commonly induced by both biotic and abiotic stress, and therefore may be important in regulating cross-talk between pathways. In chickpea, the responses to biotic and abiotic stresses were compared using a 758-probe microarray. Several shared responses were observed, in particular between high salinity stress and a fungal pathogen.

However, such studies may be of limited use in understanding plant responses to multiple simultaneous stresses, as when stresses occur in combination certain genes are activated that are not induced by either stress individually. In both *Arabidopsis* and tobacco, recent transcriptome analysis has shown that the molecular response of plants to simultaneous heat and drought stress is not additive. The stress combination results in a new pattern of gene expression that could not have been predicted by studying either stress in isolation, including the induction of numerous specifically regulated genes. Among the genes specifically regulated by the two stresses in *Arabidopsis* are those encoding heat shock proteins (HSPs), proteases, lipid biosynthesis enzymes, and starch degrading enzymes, as well as MYB TFs, protein kinases, and defence proteins involved in protection against oxidative stress.

Further transcriptome studies in various species have identified genes, hormones, and processes important in controlling plant response to multiple abiotic or multiple biotic stresses and provided targets for the improvement of stress tolerance. A combination of high light and high temperature activated a specific subset of differentially expressed genes in sunflower. In wheat, simultaneous drought and heat stress activated a pattern of HSP expression that is non-additive and could not be predicted from the study of either stress individually. At a physiological level, the imposition of drought and heat stress negatively affected wheat photosynthetic rates to a greater extent than each one alone. Analysis of the transcriptome in tobacco following infestation by two insect herbivores, a sap-feeding mirid (Tupiocoris notatus) and a chewing hornworm (Manduca sexta), revealed a specific transcriptional effect when the two herbivores were applied together compared with when each was applied separately. This supports the existence of trans-activating factors that reorganize gene expression depending on the nature of the stress. Furthermore, gene expression patterns were different if the herbivores were applied sequentially compared with in parallel, suggesting a system of priming whereby the transcriptome exhibits a long-term change following biotic attack and serves as a kind of immunological memory. Another study analysed peanut plants infected with the fungus Aspergillus parasiticus and exposed to drought. Using two expressed sequence tag (EST) cDNA libraries, 42 genes were found to be up-regulated in response to both the fungus and drought simultaneously, whereas 52 genes were up-regulated by drought alone. As root damage due to drought is advantageous for this pathogen, it is proposed that the fungus may be able to repress ABA and drought signalling in order to achieve a higher infection rate.

Transcriptome data of this kind provide a molecular explanation for the observed non-linear effects of multiple stresses on plant stress resistance described above. Rather than produce a stock response to each pathogen or abiotic stress, plants respond to the precise environmental conditions encountered and instigate a new programme of gene expression accordingly. However, despite increasing interest in the effects of simultaneous biotic and abiotic stresses on plants, no studies have yet documented whole-genome transcriptome changes resulting from any such stress combination.

Despite the quantity of data produced and the useful insights that can be inferred, transcriptomic studies do not provide a comprehensive description of changes in cellular protein activity. Research suggests that the correlation between TF and target gene mRNA is low, and that the activity of most TFs is likely to be controlled post-transcriptionally or through phosphorylation. In addition, correlation between proteomic data and transcript abundance measured by microarrays is surprisingly low (R = 0.24). In order to obtain a more accurate description of components in stress signalling pathways, metabolic and proteomic analyses are becoming increasingly utilized. The integration of these data with transcriptome results will provide the most powerful tool for characterizing such complex plant processes. In addition, the study of hormone mutants has vastly increased our understanding of multiple stress responses.

Hormone Signalling Pathways Controlling Stress Interaction

Abiotic stress responses are largely controlled by the hormone ABA, while defence against different biotic assailants is specified by antagonism between the salicylic acid (SA) and jasmonic acid (JA)/ethylene signalling pathways. However, recent findings suggest that ABA acts both synergistically and antagonistically with biotic stress signalling, creating a complex network of interacting pathways with cross-talk at different levels. ABA is therefore likely to be central in the fine-tuning of observed responses to simultaneous biotic and abiotic stresses. Treatment with ABA increases susceptibility in Arabidopsis to an avirulent P. syringae strain, in tomato to B. cinerea and Erwinia chrysanthemi, in rice to the blast fungus Magnaportha grisea, and in potato to the pathogens Phytophthora infestans and Cladosporium cucumerinum. In accordance with this, a lack of ABA can induce a higher level of pathogen resistance. For example, the ABA-insensitive Arabidopsis mutants abi1-1 and abi2-1 showed resistance to the oomycete Peronospora parasitica and the fungal pathogen Fusarium oxysporum, respectively, while the tomato sitiens mutant, which has reduced ABA levels, showed an increased accumulation of SA-dependent defence transcripts such as PR1, as well as enhanced tolerance to B. cinerea ABA treatment represses the systemic acquired resistance (SAR) pathway both upstream and downstream of SA induction in Arabidopsis and tobacco, as well as inhibiting the accumulation of crucial defence compounds such as lignins and phenylpropanoids. In turn, SA can also interfere with abiotic stress signalling. The exogenous application of SA in maize leads to drought susceptibility, while the artificial induction of SAR in Arabidopsis leads to the suppression of abiotic stress responses. In rice, resistance to the rice blast fungus M. grisea is mediated by a precise balance between ABA and SA levels. ABA also antagonizes JA and ethylene defence signalling, as shown by the ABA-mediated repression of defence genes such as PDF1.2, an effect that cannot be reversed by the application of JA or ethylene. Ethylene treatment, in turn, activates ABI1 and ABI2, two negative regulators of ABA signalling.

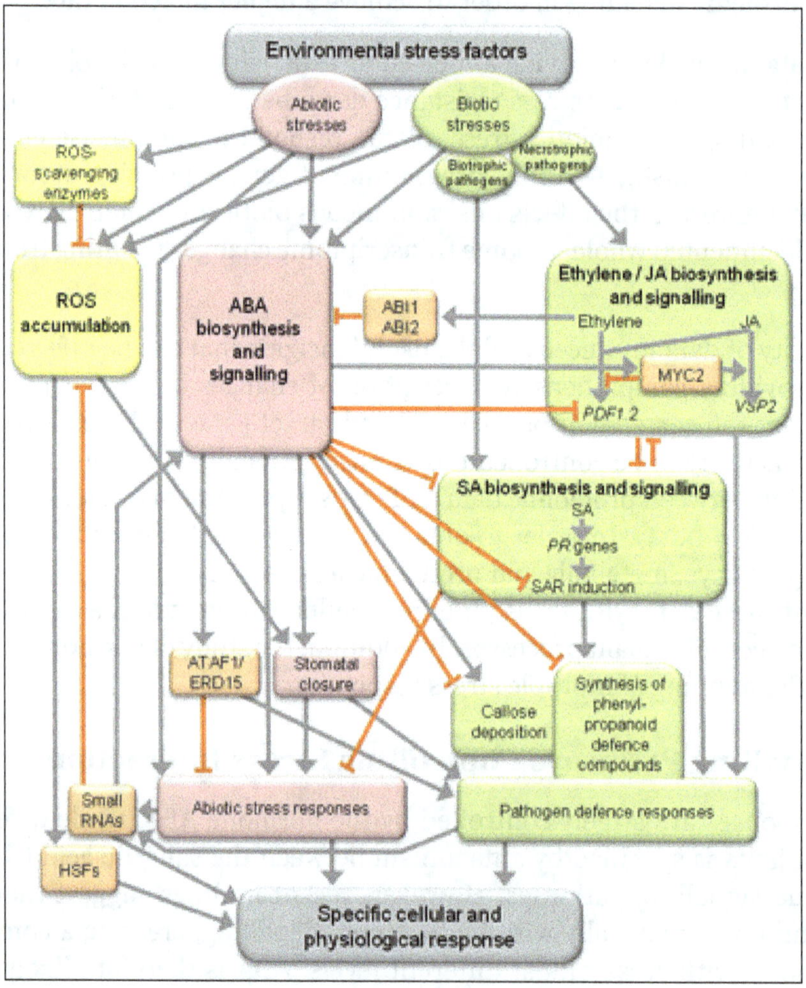

The role of plant hormones in regulating the interaction between biotic and abiotic stress. The schematic diagram shows cross-talk occurring between hormones, transcription factors, and other regulatory components when biotic and abiotic stresses occur concurrently. This complex network of interactions allows plants to respond in a highly specific fashion to the exact combination of environmental stresses encountered. Grey arrows show induction or positive regulation, while orange bars show inhibition or repression. Events characteristic of abiotic stress responses are shown in pink, while those characteristic of biotic stress responses are shown in green. Transcription factors and other regulatory genes are represented by orange boxes. ROS, reactive oxygen species; ABA, abscisic acid; JA, jasmonic acid; SA, salicylic acid; *PR*, pathogenesis-related; SAR, systemic acquired resistance; HSF, heat shock factor.

Despite these observations, ABA can also have a positive effect on pathogen defence systems. Melotto *et al.* found that ABA-induced stomatal closure was a defensive strategy of plants to prevent microbial invasion through open stomata, a process that also required intact SA signalling. Furthermore, ABA is necessary for β-aminobutyric acid (BABA)-induced callose deposition during defence against fungal pathogens, although interestingly ABA can block bacterial-induced callose production.

A new model has been proposed regarding the multifaceted role of ABA in pathogen response, whereby the influence of ABA depends on the time scale of infection and also the nature of the attacker. The model refers to three distinct phases of pathogen infection. In the first, ABA causes stomatal closure, increasing resistance to penetration by pathogens such as bacteria, and thus having a positive effect on the defence response. At this stage, ABA antagonizes SA, JA, and ethylene pathways in order to save resources, as their effects are not yet required. In the second phase, post-invasion defences centre on callose deposition to strengthen cell walls, a process that is aided by ABA during fungal infection but repressed during bacterial infection. During phase three of infection, PAMPs (pathogen-associated molecular patterns) induce the hormones SA, JA, and ethylene and long-distance signals to regulate a broad spectrum of defensive compounds. The ABA-inducible genes *ERD15* and *ATAF1* have been identified as switches that may activate ABA-dependent biotic stress responses at the expense of abiotic responses. However, increased ABA levels arising from abiotic stress conditions may repress the SA, JA, and ethylene responses even during phase three. This hypothesis provides a mechanism for the control of ABA over both biotic and abiotic stress signalling, and helps to explain previous conflicting data. To complicate matters further, recent research has found that ABA can have either a positive or a negative effect on bacteria-induced callose deposition, depending on other growth conditions such as light and glucose levels, thus switching its effect on the defence response according to the environmental situation. Such research emphasizes the need for caution when drawing conclusions from pathogen resistance assays.

ABA is now considered a global regulator of stress responses that can dominantly suppress pathogen defence pathways, thus controlling the switch in priority between the response to biotic or abiotic stress and allowing plants to respond to the most severe threat. ABA production may therefore be the crucial factor in determining how plants respond to multiple stresses.

Convergent Signalling Pathways in Multiple Stress Responses

Biotic and abiotic stress signal transduction results from a complex arrangement of interacting factors. Certain gene products are crucial to both biotic and abiotic stress signalling, and may therefore control the specificity of the response to multiple stresses. Transcriptomic and genetic analyses have greatly increased our knowledge of such processes.

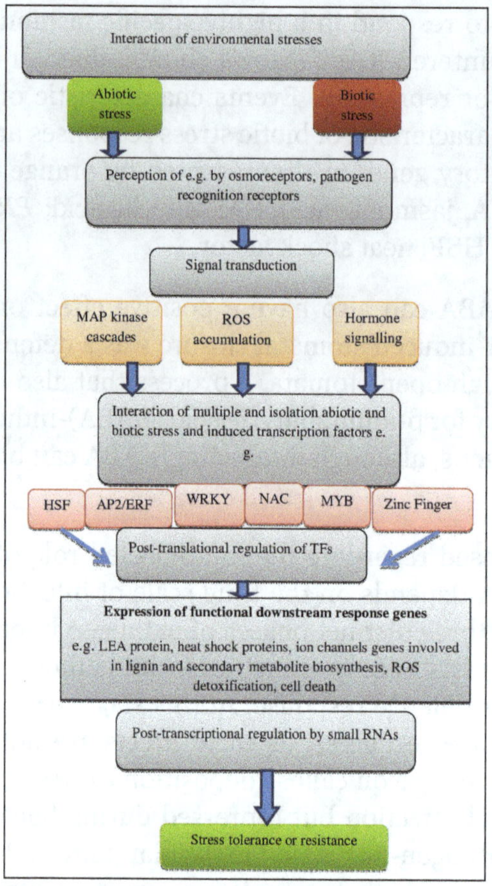

Key events in the signal transduction pathway activated in response to combined biotic and abiotic stresses.

Transcription Factors

TFs are of key importance in generating specificity in stress responses. Their manipulation provides one of the greatest opportunities for conferring multiple stress tolerance transgenically, as they control a wide range of downstream events. MYC2 (also called JIN1) is central to the interaction between biotic and abiotic signalling pathways. It is a positive regulator of specifically JA-induced defence genes but represses genes induced by combined JA/ethylene signalling. It acts as a key repressor of the SA pathway MYC2 has also been found to be activated by ABA, and myc2 mutants lack ABA-responsive gene expression. Therefore MYC2 may act as a central regulator by which ABA controls biotic stress signalling pathways. This partial synergy between ABA and JA signalling may explain the situations in which pathogen resistance is enhanced by abiotic stress, such as the finding that in barley drought stress increases resistance to Blumeria graminis, or that in Arabidopsis ABA is necessary for defence responses against the oomycete Pythium irregulare.

A large, functionally diverse group, the MYB family of TFs is becoming increasingly associated with the control of both biotic and abiotic stress responses, in particular the regulation of phenylpropanoid biosynthesis pathways. In a microarray study of combined drought and heat stress in *Arabidopsis*, certain MYB TFs were specifically induced by the stresses in combination but not by either stress individually, supporting a role for them in differentiating between precise stress conditions. MYB96 is induced by drought stress, and promotes ABA-dependent stress tolerance by inducing the biosynthesis of cuticular wax. Interestingly, MYB96 is also required for ABA-dependent SA biosynthesis and, when overexpressed, causes increased *PR* gene expression and resistance against pathogens, acting as a link between these two hormones in governing broad-spectrum stress tolerance. Antagonism between regulators of the flavonol biosynthesis pathway, MYB12 and MYB4, allows plants to manipulate the pathway depending on which stress combination is encountered. UV-B stress alone stimulates MYB12-regulated flavonol production, while the addition of the bacterial elicitor flg22 induces MYB4, repressing this process and allowing the production of defence compounds including lignin and phytoalexins. The MYB factor *BOS1* in *Arabidopsis* is also required for both biotic and abiotic stress responses. The function of MYB TFs is well conserved between species. Showing high homology to *Arabidopsis* BOS1, the tomato SlAIM1 is required for resistance to *B. cinerea* and tolerance to abiotic stresses. Its involvement in regulating ion homeostasis during oxidative stress may underpin its importance in a wide range of stresses. The MYB factor TaPIMP1 from wheat confers tolerance to pathogens, drought, and salt stresses when overexpressed in tobacco, while the rice MYB4 gene confers broad-spectrum stress tolerance when overexpressed in *Arabidopsis* or tomato. Conserved cysteine residues in the MYB family of TFs suggests that MYB activity may be controlled by cellular redox status. The oxidation or *S*-nitrosylation of cysteine residues would alter domain structure and prevent DNA binding, thus providing a mechanism for control following biotic or abiotic stress-induced reactive oxygen species (ROS) production.

NAC and AP2/ERF TFs have long been associated with stress signalling, and recent discoveries suggest that they may make excellent targets for improving broad-spectrum tolerance in crops through genetic engineering. The *Arabidopsis* NAC TF RD26 is induced by JA and pathogens, as well as by ABA, drought, and salinity, and may regulate ROS detoxification genes. There is evidence that ATAF1 in *Arabidopsis* acts as a switch to repress the effect of ABA during pathogen infection. In mutant *ataf1* plants ABA biosynthesis genes were up-regulated and resistance to penetration by the pathogen *B. graminis* was compromised. In rice, OsNAC6 is a transcriptional activator during both biotic and abiotic stresses, and confers tolerance to dehydration, high salinity, and rice blast when overexpressed, albeit with a yield penalty. The wheat *TaNAC4* gene is induced by the rust stripe fungus, high salinity, wounding, and cold stress. Its expression also responds to treatment with the hormones ABA, ethylene, and methyl jasmonate, suggesting that TaNAC4 regulates cross-signalling between stress response pathways. TSI1 from tobacco is an example of an AP2/ERF TF whose action straddles both pathogen response and abiotic stress pathways. It induces the expression of *PR* genes and can confer resistance to salt and pathogenic bacteria. Certain AP2/ERFs from tomato and soybean can enhance multiple stress tolerance when expressed in tobacco.

Other recently identified TFs that act in both pathways include tomato SlAREB1, which confers resistance to salt and drought stress as well as activating defence-related genes pepper RFP1, which functions as a defence regulator and confers osmotic stress tolerance two WRKY TFs in rice, WRKY82 and WRKY45, which are both responsive to biotic and abiotic stresses and a transcriptional repressor of DREB genes, DEAR1, which when activated represses freezing tolerance but enhances pathogen tolerance. Many TFs confer a growth impediment when overexpressed, indicative of a shift in balance between allocation of resources into stress protection rather than growth. In order to overcome this problem transgenes often need to be expressed under a stress-inducible promoter.

Mitogen-activated Protein Kinase Cascades

Mitogen-activated Protein Kinase (MAPK) cascades are crucial in eukaryotes for transducing the perception of environmental stimuli into internal signalling pathways. Plants have a particularly large number of MAPK components, allowing for the control over a wide range of stress response pathways. During biotic stress, plant transmembrane receptors such as FLS2 detect PAMPs and trigger MAPK cascades in order to establish pathogen response signalling. During abiotic stress, MAPK cascades such as the MEKK1/MKK2/MPK4/MPK6 pathway are induced, for instance following activation of the transmembrane osmoreceptor ATHK1 in *Arabidopsis*. Downstream targets of MAPK cascades include TFs, for example those in the WRKY family, and hormone response factors such as ETHYLENE INSENSITIVE3 (EIN3). MAPK cascades are particularly important in controlling cross-talk between stress responses, as many MAPKs are activated by more than one type of stress or plant hormone, and are thus able to integrate different signals. For example, *Arabidopsis* MPK6 functions in the response to ethylene synthesis, cold and salt stress, pathogen signalling, and stomatal control.

In rice at least five of the 17 identified *MAPK* genes are inducible by both biotic and abiotic stresses, and certain MAPKs can directly influence both pathways. Of these, OsMAPK5 positively regulates the response to cold, drought, and salt stress, at the same time suppressing the expression of defence transcripts such as *PR* genes. Overexpression of OsMAPK5 confers increased abiotic stress tolerance, while RNA interference (RNAi) lines are more resistant to fungal and bacterial pathogens. In tomato, the response to UV-B, a wounding signal, and a pathogen elicitor are all transduced using shared MAPK signalling components, including LeMPK1 and LeMPK2. Exposure to one stress can cause desensitization of the MAPK cascade to another stress, a phenomenon resulting from the transient refractory state of MAPKs after activation, underlining the similarity between responses. In cotton, the MAPK GhMPK16 has been shown to regulate components of biotic and abiotic signalling pathways, and when overexpressed in *Arabidopsis* confers resistance to pathogens while negatively affecting drought tolerance. MAPK cascades may be activated by and indeed regulate the production of ROS such as H_2O_2, which play a crucial role in signal cross-talk. Signal specificity of MAPK cascades is controlled by scaffold proteins such as CTR1, which can tether certain MAPK proteins together, and by spatio-temporal restrictions.

Heat Shock Factors

Heat Shock Factors (HSPs) bind and stabilize proteins that have become denatured during stress conditions, and function as molecular chaperones which prevent protein aggregation. Specific combinations of HSPs are induced following different types of abiotic stress, and they play an important role in protecting plants from oxidative stress. Recent research shows that HSPs function in the specific response of crop plants such as maize and wheat to combined heat and drought stress. The expression of HSPs is controlled by HSFs, of which there are a large number in Arabidopsis compared with animals. HSFs may act as molecular sensors to detect the presence of ROS such as H_2O_2 and activate downstream stress-responsive genes accordingly. A particular candidate for redox sensing in Arabidopsis is HSFA4a, due to its rapid induction in response to H_2O_2 and its transcriptional control over ROS-scavenging enzymes. The functional diversity between HSFs may reflect a crucial role in allowing plants to respond to different environmental conditions. In rice, the induction patterns of HSPs and HSFs following stress treatment with drought, heat, cold, and salt showed some overlap, but also a specific response to each condition. discovered that certain HSPs were specifically induced in Arabidopsis by a combination of drought and heat stress and that these changes were reflected by differences in expression levels of HSFs between stress treatments. Expression analysis of all 21 Arabidopsis HSFs revealed a different pattern for each of nine abiotic stresses, which differed depending on the tissue sampled. In addition, different biotic stresses elicited a unique combination of HSFs. The specific combination of HSFs activated may therefore direct the outcome of the stress response. Providing confirmation of this, the ectopic overexpression of the HSFA2 gene in Arabidopsis conferred tolerance to a combination of environmental stresses. The HSFA1 group of four Arabidopsis TFs (a, b, d, and e) are now thought to act as master regulators of the heat shock response. They also contribute to resistance against other abiotic stresses, and are essential for the maintenance of normal growth and productivity under stress conditions. HSFA1b regulates >500 genes, controlling tolerance to drought, extreme temperature, and a range of biotrophic pathogens, as well as being a determinant of seed yield. Activity of HSFA1b is mainly controlled post-translationally, perhaps providing an explanation for its late discovery in the study of plant stress. The targeting of this newly discovered regulatory system provides exciting possibilities for crop improvement, as when HSFA1b was overexpressed in oilseed rape it conferred stress resistance phenotypes as well as maintaining yield productivity.

Reactive Oxygen Species

There is increasing evidence that ROS play a central role in mediation between biotic and abiotic stress responses. ROS are critical to both types of stress response, but function differently in each. Low levels of ROS accumulate constantly as by-products of metabolism, but their concentration increases during abiotic stress conditions such as osmotic stress, salinity, and high light. To minimize damage caused by these potentially harmful molecules, plants produce antioxidants and ROS-scavenging enzymes. In contrast, plants actively generate ROS following pathogen infection, in a process known as the oxidative burst. This limits pathogen spread by contributing to the hypersensitive response and cell death, in this case requiring the coordinated down-regulation of ROS-scavenging mechanisms.

Plants have also adapted to use ROS as stress signal transduction molecules, and a unique footprint of ROS-responsive genes is induced by each type of biotic and abiotic stress. H_2O_2 is produced by membrane-bound NADPH oxidases immediately on pathogen infection or wounding, then diffuses into cells and activates various plant defences, for example in distal leaf tissue following wounding. ROS production is furthermore required for ABA-driven stomatal closure. ABA induces the NADPH oxidases *AtRbohD* and *AtRbohF* which generate ROS in guard cells as signalling intermediaries. This commonality between biotic and abiotic stress-induced ROS production may contribute towards the positive effect of ABA on early pathogen response.

However, ABA can also have a negative effect on ROS-mediated pathogen defence, as demonstrated by the ABA-deficient tomato *sitiens* mutant. These plants accumulate H_2O_2 rapidly following infection by *B. cinerea*, leading to a stronger defence response and greater resistance to the pathogen compared with the wild type. The ABA biosynthesis pathway is tightly linked to ascorbate levels and therefore to the redox status of cells. It has been proposed that the build-up of ABA precursors in the *sitiens* mutant could disrupt redox homeostasis and cause the observed beneficial increase in ROS accumulation.

Precise coordination of ROS production and scavenging mechanisms during combined biotic and abiotic stress is clearly necessary, and certain TFs integrate ROS-scavenging mechanisms in response to multiple types of stress. ZAT12 is an Arabidopsis zinc-finger TF that is responsive to high H_2O_2 levels and is induced by wounding, and abiotic and biotic stress. It induces genes such as ascorbate peroxidase (APX1), and when overexpressed conferred tolerance to oxidative stress, freezing, and high light. Similarly, ZAT7 is activated in response to oxidative stress, and confers tolerance to salt, cold, and oxidative stress when expressed ectopically. ZAT7 also induces the expression of various defence transcripts, including WRKY70, a gene that modulates cross-talk between the SA and JA pathogen response pathways. Evidence suggests that the APX1 enzyme itself is crucial in the response to multiple stresses. It was found to accumulate in Arabidopsis specifically in response to combined heat and drought stress, but not in response to either stress individually.

Plants lacking the APX1 gene have a significantly lower survival rate under the stress combination compared with wild-type plants, although survival under a single stress is unaffected. The cytosolic location of APX1 suggests that dual heat and drought stress may cause a critical build-up of H_2O_2 in the cytosol, leading to cell injury and death if not dealt with. The cellular localization of ROS production and metabolism may determine specificity in ROS signalling, as different scavenging enzymes and antioxidants are prevalent in each cellular compartment. A newly discovered gene, pentatricopeptide repeat protein for germination on NaCl (PGN), highlights the role of ROS in the interaction between biotic and abiotic stress responses. There is evidence that PGN regulates ROS homeostasis in the mitochondria by interacting with genes such as alternative oxidase 1 (AOX1). In the pgn mutant, impaired ROS regulation caused susceptibility to necrotrophic fungal pathogens, high salt, and glucose stress, and led to increased levels of ROS and ABA. Overexpression of PGN led to similar stress susceptibility phenotypes as in pgn plants, indicating that tight regulation of PGN is necessary for the control of mitochondrial ROS levels in both biotic and abiotic stresses.

As stresses often occur in combination, and the relationship between ROS signalling mechanisms in different stress responses is complex, identifying master regulators of ROS metabolism may provide the best candidates for generating broad-spectrum stress tolerance.

Small RNAs

Our understanding of the complexity of plants' responses to stress has been enhanced by the discovery of small non-coding RNA species that play crucial regulatory roles. The levels of these small RNAs change in response to different abiotic stresses in Arabidopsis, providing new avenues for the investigation of plant stress signalling. MicroRNAs (miRNAs), which are derived from long single-stranded RNA folded into a hairpin structure, and small interfering RNAs (siRNAs), which are derived from double-stranded RNA and can act in both cis and trans, silence genes post-transcriptionally by targeting mRNA for degradation or repressing translation. The advent of high-throughput sequencing technology has facilitated rapid identification of small RNAs, and several categories are now known to be involved in stress responses, as reviewed by Ruiz-Ferrer and Voinnet, Sunkar et al, and Mallory and Vaucheret. For example, in Arabidopsis, the miR399 miRNA is induced in response to low phosphate stress. It controls phosphate homeostasis by suppressing the expression of a ubiquitin-conjugating enzyme, thus allowing uptake and mobilization of phosphate and preventing phosphate starvation. Although not generally well conserved between species, stress-responsive miRNAs have been identified in a variety of plants.

In the grass Brachypodium distachyon, the induction and suppression of miRNAs occur as a result of cold stress. In rice a network of oxidative stress-responsive miRNAs regulate processes such as programmed cell death, nutrient transport, and auxin homeostasis, while the siRNAs siR441 and siR446 positively regulate ABA signalling and tolerance to abiotic stress. In the legume Medicago truncatula the drought-induced production of MtDCL1 and MtAGO1, enzymes that regulate the production and processing of miRNAs, provides a mechanism for the regulation of these small RNA species. Improvement of stress tolerance by manipulation of miRNA regulation has also been demonstrated. Transgenic Arabidopsis plants expressing a form of the chloroplastic superoxide dismutase CSD2 that is resistant to suppression by mi398 showed higher ability to detoxify superoxide radicals and thus resistance to oxidative stress.

The ability of miRNAs to modulate events in ROS regulation may explain their function in both biotic and abiotic stress response pathways. The expression of 11 soybean miRNAs was investigated in response to water deficit and infection with soybean rust fungus. All miRNAs were differentially regulated by each stress, usually in an opposite direction. One miRNA, MIR-Seq11, targets a peroxidase protein that is involved in metabolism of H_2O_2 during oxidative stress, potentially inhibiting its translation or causing mRNA decay. The down-regulation of MIR-Seq11 during biotic stress may therefore provide protection from ROS during pathogen infection. Differential expression between drought-tolerant and drought-susceptible cultivars was also observed. MIR-Seq11 was up-regulated in the drought-susceptible cultivar but not in those which were drought tolerant, suggesting that miRNA control of ROS scavenging could also be a factor in determining susceptibility of these plants to drought. In wheat, analysis of miRNA expression in response to heat stress and infection with the powdery mildew fungus *Erysiphe graminis* revealed that members of the same miRNA family were differentially regulated by the stresses. Some were induced by heat but suppressed by the pathogen, while others were induced by both stresses. The unique pattern of miRNA production may therefore provide specificity in controlling stress signalling.

A newly discovered class of siRNAs called nat-siRNAs can regulate both biotic and abiotic stress responses. These are derived from natural antisense genes that overlap on opposite strands in the genome, and are formed by the annealing of the sense and antisense transcripts. For example nat-siRNAs derived from the antisense gene pair PC5DH–SRO5 control proline levels and thus tolerance to salt in Arabidopsis. In response to salt stress, SRO5 is transcriptionally activated and a corresponding nat-siRNA is produced, which targets the mRNA of the proline degradation gene PC5DH for destruction. In response to recognition of the P. syringae effector AvrPt2, a nat-siRNA called nat-siRNAATGB2 is specifically induced, which has the effect of down-regulating a putative constitutively active negative regulator of defence.

The control of gene expression by small RNAs may provide a crucial mechanism for regulating the growth and reproduction of plants experiencing stress conditions. This hypothesis is supported by the observation that many stress-responsive small RNAs are also involved in developmental processes, and that disruption of miRNA signalling in Arabidopsis can result in delayed flowering time, reduced fertility, leaf defects, and a reduced number of lateral roots.

Controlling Specificity in Molecular Stress Interactions

The complex control of specificity in stress signalling is enhanced by a variety of additional cellular interactions. TFs do not bind indiscriminately to every gene with a matching response element, as there may be thousands of such sites across the genome. Instead, specificity can be generated by TFs forming into homo- or heterodimers, which then bind to a pair of response elements at an appropriate distance from each other, or by the the co-operation of bridging or scaffold proteins which direct TFs to the correct response elements. In addition, TFs may be regulated themselves through redox modification of cysteine residues that can prevent transport into the nucleus, phosphorylation, or protein degradation. These systems allow eukaryotes an extremely high level of control over cellular processes, generating a large number of downstream signalling options without having to produce such high numbers of TFs in the nucleus that overcrowding occurs and TFs cannot easily associate with their target sequence. One mechanism for enabling a rapid TF response is that of proteolytic activation, in which membrane-bound TFs are cleaved by stress-responsive proteases allowing re-localization of the active subunit into the nucleus. For example, the membrane-bound NTL6 TF becomes cleaved in response to cold stress, and travels to the nucleus where it activates cold-responsive *PR* genes, integrating the responses to cold and pathogens. The overexpression of *NTL6* in *Arabidopsis* leads to pathogen resistance, while plants with reduced NTL6 activity are susceptible to pathogens at cold temperatures. This illustrates a means by which plants protect themselves from pathogens during cold times of year, a system also thought to occur in grass species such as winter rye. Transcriptional activation of response elements also requires chromatin remodelling, a process that can be achieved through stress-induced histone modification. Plant mechanisms for increasing signalling specificity have been recently reviewed by Vaahtera and Brosche.

Perspectives: Seeing Plant Stress Research in a New Light

Hundreds of studies investigating the effect of individual and combinatorial stresses have allowed us to piece together the complex network of molecular interactions controlling plant stress responses. Plants activate both specific and non-specific stress responses as a reaction to adverse environmental conditions, allowing them to maximize efficiency in responding to the exact set of conditions encountered, at the same time as conserving resources for growth. Signal specificity is achieved through the precise interplay between components of each pathway, particularly the hormones ABA, SA, and JA, TFs, HSFs, ROS, and small RNAs. With so many interacting factors it is easy to see how the effect of multiple stresses on an individual resistance response can be unpredictable. Traditionally, individual plant stress factors have been studied as isolated stimuli that trigger linear signalling pathways. It is clear that this model is no longer sufficient, and both biotic and abiotic stress pathways are inextricably linked in a broad network of molecular interactions.

One of the ultimate aims of plant stress research is to provide targets for the improvement of stress tolerance in crop plants. With the forecast changes in climatic conditions, novel stress combinations are likely to occur in agricultural systems in many parts of the world, presenting new challenges in creating multiple stress-tolerant crops. The development of such plants will depend on understanding crucial stress-regulatory networks and the potential effects of different combinations of adverse conditions. Studies of multiple stress responses in *Arabidopsis* and other species have revealed several possible avenues for exploration. Common threads have been identified between biotic and abiotic signalling pathways, which may lead to their antagonistic nature. Master regulatory genes that act in both biotic and abiotic stress response systems are excellent candidates for manipulating stress tolerance. HSFs such as those in the HSFA1 and HSFA2 families control a large number of downstream stress-responsive genes and have shown promise in conferring tolerance to multiple stresses. Members of the MYC, MYB, and NAC TF families are also potential targets, as they are involved in controlling antagonism between hormone-mediated abiotic stress and pathogen response pathways. Research has already shown that their manipulation can confer tolerance to biotic and abiotic stresses in several species. Controlling ROS homeostasis genes may also be able to affect tolerance to a wide range of stresses, due to the involvement of ROS in both abiotic stress and pathogen stress.

The pyramiding of transgenes has proven successful in conferring tolerance to a particular stress, for example to drought in wheat, to salt stress in tobacco, to nematodes in potato, and to insects in chickpea. Genetically modified crops have been approved with up to three pyramided transgenes, conferring resistance to two different biotic stresses, for example a virus- and insect-resistant potato developed by Monsanto, and the area of land cultivated with crops containing pyramided transgenes is rising steadily. The pyramiding of several genes may therefore provide further opportunities for creating broad-spectrum stress-tolerant plants.

Another lesson that can be learnt from the extensive study of plant stress responses is that experiments should be carried out in conditions that relate closely to natural or field conditions. Plants' efficiency in sensing and responding to each unique set of environmental conditions means that different methods of imposing stress can lead to drastically different transcriptional profiles. Discrepancies between results of similar experiments show that changes in growth parameters such as light or sugar levels can significantly affect plant stress response priorities. Therefore, in order to provide a model for crop stress responses, an integrated approach should be adopted whereby researchers aim to mimic a field stress situation in terms of growth media, duration, and severity of stress, timing of stress, and nutrient availability. In addition, it is now clear that plant varieties that have been developed with improved stress tolerance should be tested under the full range of stress combinations that are likely to occur in the field, rather than each stress individually.

In recent years, the necessity for a change in focus in plant abiotic stress research has become apparent. As plants have finite resources which must be balanced between growth and defence against stresses, often both natural and induced stress tolerance comes with a growth or yield penalty, making it agriculturally disadvantageous. Rather than developing crops that can survive extreme stress events, it may therefore be more beneficial to focus on producing crops which are stress tolerant but which maintain high photosynthesis, growth rates, and yield. The C24 genotype of Arabidopsis exhibits such traits. This accession exhibits constitutive expression of SA-induced defences which allow high levels of resistance against P. syringae and Hyaloperonospora arabidopsidis, but, unlike mutant genotypes with constitutive SA responses, C24 suffers no yield loss. It also has a greater water use efficiency, meaning that it requires less water than other accessions to provide the same seed yield, and is drought tolerant. This finding suggests that it may be possible to create plants with broad-spectrum stress tolerance without affecting yield, and continuing studies are being carried out to determine the genetic basis for this beneficial trait. The homeostasis of energy levels also provides a link between stress responses and maintenance of growth. Energy deprivation in plants results from most types of stress as well as nutrient starvation and darkness. The Arabidopsis protein kinase SnRK1 acts as a sensor for energy depletion, and restores homeostasis through large-scale transcriptional re-programming, thus enabling the continuation of growth and metabolism. Amongst its targets are a wide range of general stress-responsive genes involved in both pathogen and abiotic stress tolerance. The manipulation of energy sensors such as SnRK1 could therefore play an important role in maintaining photosynthesis and yield during stress conditions. Targeting of miRNAs that act both in stress responses and in regulation of growth and development may provide further possibilities. The challenge for plant scientists in the 21st century will be to develop stable multiple stress tolerance traits in agronomically important crop plants, thus improving yields particularly in areas with adverse environmental conditions, and contributing to global food security.

AGRICULTURE BIOTECHNOLOGY FOR BIOTIC AND ABIOTIC STRESS

Biotechnology and Stress

Biotechnology nowadays changing the agricultural and plant scene in three major areas: growth and development control (vegetative and reproduction/propagation), protecting plants against the ever-increasing threats of abiotic and biotic stress, expanding the horizons by producing specialty foods, biochemicals and pharmaceuticals. Multiple biotic and abiotic environmental stress factors affect negatively various aspects of plant growth, development, and crop productivity. Plants, as sessile organisms, have developed, in the course of their evolution, efficient strategies of response to avoid, tolerate, or adapt to different types of stress situations. The diverse stress factors that plants have to face often activate similar cell signaling pathways and cellular responses, such as the production of stress proteins, upregulation of the antioxidant machinery, and accumulation of compatible solutes. Over the last few decades advances in plant physiology, genetics, and molecular biology have greatly improved our understanding of plant responses to abiotic stress conditions.

Biotechnology and Sustainable Agriculture

Biotechnology has been contributing to sustainable agriculture through the following ways:

- Increased resistance against biotic stresses (insect pests and diseases);
- Increased resistance against abiotic stresses (drought, cold, flooding, and problem soils);
- Bioremediation of polluted soils and biodetectors for monitoring pollution;
- Increased productivity and quality;
- Enhanced nitrogen fixation and increased nutrient uptake and use efficiency;
- Improved fermentation technology;
- Improved technologies for generating biomass-derived energy;
- Generation of high nutrient levels in nutrient-deficient staple crops such as rice.

The main biotechnological approaches to improve plant responses to stress are:

Breeding

Climate change threatens sustainable agriculture with its rapid and unpredictable effects, making it particularly difficult for agricultural scientists and farmers to respond to challenges from biotic and abiotic stresses. Global warming is causing changes in temperature at a rate unmatched by any temperature change over the last 50 million years making it all the more important how well agricultural production can be maintained. It calls for concerted efforts in sustaining food production to meet this challenge. Aspects of climate change that may affect future crop production include changes in both mean and extreme temperatures and changes to available water. Improving adaptation requires development of plants and associated production and management systems to cope with or avoid climate effects. Many complex processes and interactions determine crop yield under climate change and crop simulation models combined with high resolution climate change scenarios may identify key traits that are important under drought and high temperature stress in crops. Climate change and associated strategies should build over adopting new technologies and farming policy and systems. Drought and high temperatures are considered as key stress factors where scientists should anticipate the effects of climate change on plants. With the knowledge and tools, plant breeders and biotechnologists that will begin to make the best use of the huge global effort to meet challenges posed by climate change.

Drought and soil salinity are major abiotic stress factors affecting seriously crop production and food safety. They reveal adverse impacts on the socio-economic structure of many developing countries. Water scarcity, declining water quality for irrigation, and soil salinity are problems which are becoming really acute. It was estimated that 20% of all cultivated land and nearly a half of irrigated land are affected by salt, greatly reducing the yield of crops below their genetic potential. Breeding cultivars that combine drought and temperature resistance, yield potential, and yield stability are a prerequisite for stable productivity will require introgressing resistance genes from landraces and wild relatives to commercial cultivars and evaluating them in a matrix of stress environments. Achieving genetic increases in yield under abiotic stresses has always been a difficult challenge for plant breeders.

The accumulation of soluble salts in soil leads to an increase in the osmotic pressure of soil solution, which may limit the absorption of water by the seeds or by the plant roots. Salt damage to plants is attributed to the reduction in water availability toxicity or specific ions, and nutritional imbalance caused by such ions. There has been an active debate in literature whether to breed primarily for yield potential or for improved yield under the stressful environments that prevail in most wheat-growing areas worldwide. One can find examples of both views: (i) improving yield potential constitutively improves yield under stress conditions; and (ii) breeding for yield potential produces lines of poorer behavior than landraces or lines selected for better performance under stress conditions. As the human population grows, the demand for animal products increases, thus requiring improved agriculture systems. Increases in crop yields have boosted the demand for human feed. To feed the increasing human population, more land will need to be devoted to crops, thereby reducing the land available for pasture and fodder. The limited land available for feed and fodder production and the decreasing quality of available pasture have given rise to the use of improved crop genotypes through breeding systems for increasing human demand for food makes feed improvement essential in order to avoid competition between animal and human feed requirements; feed improvement and use in developing targeted sites should therefore be based on farmer practices, production systems and participation; locally available and potential feed resources production systems and improved techniques appropriate to the farming system in use and economically and socially acceptable. Wheat and barley have evolved as grain structures, destined more to be double purpose grain crops rather than forage crops. These crops have high water-use efficiency and are well adapted to dry areas (abiotic stress). For instance, these crops also have a high degree of drought, salinity and heat tolerance. Adaptation of these crops to soil problems and degraded lands further broadens their adaptation range. Elite breeding lines, open-pollinated varieties and hybrid parents have been identified in wheat, barley through classical breeding programs and selection methods for the most tolerant genotypes and that have performed well at high salinity levels (10dS m-1).

Biotechnology has many potential uses in the campaign against pests and diseases, providing knowledge and tools useful to the plant breeder. However, the very complex nature of the relationships between pests, diseases, vectors, host plants and the environment means that integrated management methods are required. Biotechnology is certainly one component of an integrated management strategy along with resistant varieties, biocontrol, appropriate cultural practices and rational agrochemical use.

Host plant resistance is an important tool to control diseases of major food crops in developing countries, especially wheat, rice, potato, cassava, chickpea, peanuts and cowpea. The use of resistant varieties is very much welcomed by resource poor farmers because it does not require additional cost and it is environment-friendly. Rice varieties resistant to rice blast, bacterial blight, and brown spot are widely used. Such success stimulated interest in extending the principles of genetic diversity for disease control to other crops in developing countries. Rusts have been known to cause serious disease on wheat since its domestication. The use of genetic resistance is still the most economic and feasible mode of disease control. Genetic resistance is often based on a limited number of major genes that are readily overcome by evolving pathogen races. With the reduction of genetic diversity in the wheat cultivars planted over large areas globally, serious rust epidemics are being recorded whenever new aggressive virulent rust races emerge. A typical example is the yellow rust epidemics that spread from East Africa to Central and South Asia and North Africa during the 1980's and 1990's. Presently the breakdown of Yr27, a gene used to replace Yr9, and the emerging stem rust race Ug99 are threatening 80-90% of commercial wheat varieties grown worldwide.

Genomics

A gene by gene approach has been typically used to understand its function. Functional genomics allows large-scale gene function analysis with high throughput technology and incorporates interaction of gene products at cellular and organism level. Gene identification through physical and chemical mutagens has become amenable for large-scale analysis with the availability of markers, but gene tagging is more promising for functional analysis on a wider scale. Moreover, the understanding of the complexity of stress signaling and plant adaptive processes would require the analysis of the function of numerous genes involved in stress response. Numerous investigations show that plant defense response genes are transcriptionally activated by pathogens and also by different types of abiotic stress. It has been described that the induction of specific defense genes, in the response against certain pathogens, is dependent on specific environmental conditions, suggesting the existence of a complex signaling network that allows the plant to recognize and protect itself against pathogens and environmental stress. Advances in plant genomics research have opened up new perspectives and opportunities for improving crop plants and their productivity. The genomics technologies have been found useful in deciphering the multigenicity of biotic and abiotic plant stress responses through genome sequences, stress-specific cell and tissue transcript collections, protein and metabolite profiles and their dynamic changes, protein interactions, and mutant screens.

Proteomics

The adaptation of plants to biotic or abiotic stress conditions is mediated through deep changes in gene expression which result in changes in composition of plant transcriptome, proteome, and metabolome. Since proteins are directly involved in plant stress response, proteomics studies can significantly contribute to elucidate the possible relationships between protein abundance and plant stress acclimation. Several studies have already proven that the changes in gene expression at transcript level do not often correspond with the changes at protein level. The investigation of changes in plant proteome is highly important since proteins, unlike transcripts, are direct effectors of plant stress response.

Proteomics studies could thus lead to identification of potential protein markers whose changes in abundance can be associated with quantitative changes in some physiological parameters related to stress tolerance.

Metabolomics

The possibility of monitoring a complete set of metabolites could largely improve the understanding of many physiological plant processes. It is systematic study, defined as "metabolomics," is intended to provide an integrated view of the functional status of an organism. Besides its use as a breeding or selection tool, metabolomics techniques have also been used to evaluate stress responses in barley, Citrus and Arabido psisthaliana. Some of the metabolites that have been involved in the plant responses to stress.

Crop Genetic Improvement

Use of modern molecular biology tools for elucidating the control mechanisms of stress tolerance and for engineering stress tolerant plants is based on the expression of specific stress-related genes. To date, successes in genetic improvement of environmental stress resistance have involved manipulation of a single or a few genes involved in signaling/regulatory pathways or that encode enzymes involved in these pathways. The plant hormone abscisic acid (ABA) regulates the adaptive response of plants to environmental stresses such as drought, salinity, and chilling via diverse physiological and developmental processes. The ABA biosynthetic pathway has been deeply studied, and many of the key enzymes involved in ABA synthesis have been used in transgenic plants to improve abiotic stress tolerance. Transgenic plants overexpressing the genes involved in ABA synthesis showed increased tolerance to drought and salinity stress.

A variety of insects, mites and nematodes significantly reduce the yield and quality of the crop plants. The conventional method is to use synthetic pesticides, which also have severe effects on human health and environment. The transgenic technology uses an innovative and eco-friendly method to improve pest control management. About 40 genes obtained from microorganisms of higher plants and animals have been used to provide insect resistance in crop plants.

The success of the transgenic approach led to the development of Bt crops, transgenic crops are used worldwide to control major pests of cotton, corn and soybean. Cotton (Gossypium hirsutum) tolerant to lepidopteran larvae (caterpillars), maize (Zea mays) tolerant to both lepidopteran and coleopteran larvae (rootworms) and soya bean (Glycine max) both lepidopteran and coleopteran larvae have become widely used in global agriculture and have led to reductions in pesticide usage and lower production costs.

Elicitors for Management of Abiotic and Biotic Stresses

Plant environmental stress such as drought conditions, high water or soil salinity or too cold or too hot temperatures represents the most important economic problem for crop production worldwide. In plants, abiotic stress is often accompanied by an excess in ROS levels that leads to oxidative damage. Thus, plant tolerance to abiotic stresses involves adaptive changes in plant morphology, physiological and biochemical processes to minimize stress-induced oxidative injury. ROS play a dual role in plant stress response: they are toxic by-products that accumulate in cells but they are also important signal transduction molecules. The signal perception of an abiotic stress by the plasma membrane is followed by the generation of second messengers such as calcium, ROS and inositol phosphates. These messengers modulate the calcium level in the cells. This change is recognized by calcium sensors, resulting in the expression of major stress responsive genes, and finally leading to a physiological response. Certain chemical treatments can induce an increase of plant stress responses in monocotyledonous plants. Application of secondary messengers such as calcium or inositol to plants enhances their tolerance towards abiotic stress. Seed priming with $CaCl_2$ reduced chilling injury of maize. Plant growth was improved, antioxidant activity was enhanced and soluble sugars accumulated to a higher level after calcium treatment. Ascorbic acid pretreatment of sugarcane results also in higher salt tolerance triggered by enhanced antioxidant enzymes activities. It is now established that polyamine biosynthesis can inhibit the growth of a wide range of fungi by exerting fungicidal activity, thereby reducing the infection of a range of plant pathogenic fungi. Polyamines are essential for normal growth and development, the regulation of several cellular and molecular functions, as well as during the plant's response to stress. Polyamine levels are also known to change in plants in response to biotic stress in the form of pathogen infection. Antioxidants are known to play an important role in the resistance of plants against pathogen attack.

Bio-elicitors for Management of Abiotic and Biotic Stresses in Crop

Recently, great attention has been devoted to cultivate field crops in new reclaimed sandy soils. In general, under such unfavorable conditions and in soil characterized as low fertile, low organic matter content and high leaching rate thus the production of most crops is not economic and farmers have to apply high rates of chemical fertilizers to maintain satisfactory yield. Improvements in its yield have been established due to a large number of factors. Soil microbial populations are immersed in a framework of interactions, which are known to affect plant fitness and soil quality. Beneficial free-living bacteria in the plant rhizosphere are usually referred as plant growth promoting rhizobacteria (PGPR) which promote the plant growth by means of direct and indirect mechanisms. To rescue plant growth in such stressful conditions, PGPR have been known to play an essential role in the growth and metabolism of plants. Many PGPR contain the enzyme 1-aminocyclopropane -1- carboxylate (ACC) deaminase and promote plant growth by sequestering and cleaving plant produced ACC, the immediate precursor of the plant hormone ethylene and thereby lowering the level of ethylene in the plant. Some microorganisms and the molecules they produce are able to biocontrol plant pathogens by inducing SAR and thus can be defined as biocontrol microorganisms(BCMs). Some fungal BCMs are able to promote plant growth and development, so acting as PGPMs, that in turn determines a higher tolerance of the plants against abiotic stresses, such as drought and salinity. Both BCMs and PGPMs can be defined as "biostimulant microorganisms", able to foster plant growth and defense against pathogens throughout the crop life cycle, from seed germination to plant maturity. Indeed, BCMs, whose main action is to prevent or inhibit the growth of pathogens by SAR, exercise" indirect" benefits on plant growth by antibiosis based on the production of hydrolytic enzymes or inhibiting substances. These indirect effects have been clarified only in part, and even less is known regarding the "direct" effects of BCMs on the improvement of plant growth through production of siderophores and phytochelatins, which chelate metals and make them available to the roots.

The most interesting PGPMs are those able to colonise the rhizosphere. This latter is particularly richin nutrients and supports a microbial population that can exert positive effects on the physiological state of the roots, on the absorption of nutrients and on plant tolerance to environmental stresses. A wide range of abiotic stresses such as flooding, drought, salt, heavy metals and organic contaminants; high and low temperature can induce synthesis of stress ethylene which causes severe damage to the plants and affecting the crop production. ACC deaminase containing PGPR can be exploited as successful strategy for protecting the plants against the deleterious effects caused by these stresses. Amongst the biotic stresses, phytopathogens can reduce crop yield which is an enormous potential loss to crop productivity. Development of superior or novel PGPR strains by improving above traits can be possible using genetic manipulations. These PGPR-biotechnologies can be exploited as a low-input, sustainable and environment-friendly technology for the management of plant stresses. PGPR are able to increase plant tolerance via different mechanisms such as lowering ethylene concentration in plants, producing phytohormones, regulating nutrient uptake, inducing and augmenting stress-response gene expression or the production of antioxidants.

Beneficial effects of drought tolerant Pseudomonas strains on drought-stressed maize plants were observed at the morphological and physiological level. Interestingly, antioxidant enzyme activities were lower in bacteria-inoculated plants compared to control plants. This shows that the biochemical response of inoculated plants correspond to a less stressed plant. In wheat, salt-tolerant rhizobacteria induced phenolics and quercetin accumulation leading to an enhancement of plant growth under saline stress. PGPR containing ACC-deaminase can improve maize plant growth under salt stress conditions through better nutrient uptake and exopolysaccharides produced by PGPR allow maize plants to tolerate salt stress by binding Na^+, resulting in a reduced salt uptake by the plant. The bacteria belonging to Bacillus spp. are ubiquitous microorganisms, present inthe soil and in the phylloplane, and they are also able to live as endophytes. They have been studied for their antagonistic activity and induction of plant resistance against stresses. endophyte isolates of B. subtilis, able to control several diseases caused by leaf and soil pathogens, have been identified. Many Bacillus isolates can promote plant vegetative development by producing several extracellular substances, so acting as PGPMs. Paenibacillus polymyxa, a common soil bacterium, belongs to this group. A range of activities has been found to be associated with P. polymyxa treatment, some of which might be involved in plant growth promotion.

Most mechanisms proposed to explain indirect growth promotion suggest that the active principle may be a secondary bacterial metabolite that antagonizes pathogens as HCN, side rophores, and antibiotics. P. polymyxais known to produce antibiotic compounds, and inoculation with P. polymyxa suppresses several plant pathogens. Inoculation by the PGPR P. polymyxa can protect A. thaliana against a bacterial pathogen and drought stress in a gnotobiotic system. This effect correlates with an increase in the expression not only of genes associated with biotic stress (PR-1, HEL, ATVSP) but also of those associated with drought stress (ERD15, RAB18).

In recent years, the induction of the plant defense response by fungi and bacteria that normally colonize living plants without causing visual damage, has received ample attention. Endophytic microorganisms that reside in the intercellular spaces of higher plants can also induce the plant defense response. In some cases, endophytes directly accelerate seedling emergence, promote plant establishment under adverse conditions and enhance plant growth and development. There are many examples of compounds produced as a result of endophyte infection which accumulate to higher levels than normal. In Hordeum, both proline and ergot alkaloids have been identified. Microbial elicitors derived from some fungal endophytes promote biomass and induce terpenoid biosynthesis and production in plant suspension cells. As sustainable and renewable agricultural production increases in prominence, endophytic microorganisms will increasingly play important roles and offer environmentally-friendly methods to increase productivity while reducing chemical inputs. A particular attention should be given to endomycorrhizal (Glomus spp.) and rhizosphere (Trichoderma spp.) coloniser that could allow plants to achieve optimum yields. The SAR represents a valid opportunity in plant natural protection. Therefore, the research activities should be oriented to the use of BCMs as inducers of SAR in agronomically important species against some of their most severe pathogens. Trichoderma spp. and Glomusspp. are some of the most abundant fungi found in many soil types, able to colonise plant roots and plant debris. They are agriculturally and industrially important, being the major source of manycommercial biostimulants and biofungicides. On the contrary, many Trichoderma and Glomus species are strong BCMs against bacteria, fungi and nematodes, and for this reason more than 60 % of all registered biostimulants used for plant disease control are Trichoderma- and Glomus–based.

Trichoderma spp. are also important for their ability to synthesisepeptaibols, a family of peptides with antibiotic function. Their antibiotic functionarises from their membrane-insertion and pore-forming abilities, and it has been shown that peptaibols produced by T. pseudokoningii can induce programmed cell death in plant fungal pathogens.

Moreover,the crosstalk between the different plant hormones, whose levels change after plantinoculation with PGPMs, results in synergetic or antagonistic interactions that play crucial roles in response of plants to abiotic stress, such as drought, salinity and toxic metals. Thus, plant hormones play central roles in the ability of plants to adapt changing environments by mediating growth, development, nutrient allocation and source/sink transitions.

PGPR-nematodes Interactions

The PGPR-nematodes interactions have been extensively studied with the aim to manage plant-parasitic nematodes. These studies involve the selection of bacteria that can be used as biocontrol agents against nematodes. The genera involved include Agrobacterium, Alcaligenes, Bacillus,

Clostridium, Desulfovibrio, Pseudomonas, Serratiaand Streptomyces. These bacteria were characterized for production of hydrolytic enzymes, HCN, phenol oxidation and antifungal activity.

Organic fertilizers are safe for both the environment and human health. It is used to improve soil properties making the soil easier to cultivate by encouraging root development, providing plant nutrients and enabling their increased uptake by plants. Abou-Aly and Gomaa found that dual inoculation of coriander seeds with Azotobacter and phosphate solubilizers increased the vegetative growth and photosynthetic characteristics.

References

- Spoel, Steven H.; Dong, Xinnian (2012). "How do plants achieve immunity? Defence without specialized immune cells". Nature Reviews Immunology. 12 (2): 89–100. Doi:10.1038/nri3141. PMID 22273771
- Agriculture-biotechnology-for-management-of--multiple-biotic-and-abiotic-environmental-stress-in-crops: jocpr.com, Retrieved 25 May, 2019
- Taiz Lincoln, Zeiger Eduardo, Møller Ian Max, Murphy Angus (2015). Plant Physiology and Development. USA: Sinauer Associations, Inc. P. 706. ISBN 9781605352558
- Van Huijsduijnen, R. A. M. H.; Alblas, S. W.; De Rijk, R. H.; Bol, J. F. (1986). "Induction by Salicylic Acid of Pathogenesis-related Proteins and Resistance to Alfalfa Mosaic Virus Infection in Various Plant Species". Journal of General Virology. 67 (10): 2135–2143. Doi:10.1099/0022-1317-67-10-2135

4

Pest and Weed Management

The approach to pest and weed regulation which involves increasing the strengths of natural systems to reinforce the natural processes is referred to as ecological pest and weed management. It involves the use of bioeffectors, bioherbicides and biopesticides. These aspects of ecological pest management have been thoroughly discussed in this chapter.

ECOLOGICAL PEST MANAGEMENT

Ecological Pest Management (EPM) is an approach to increasing the strengths of natural systems to reinforce the natural processes of pest regulation and improve agricultural production. Also know as Integrated Pest Management (IPM), this practice can be "defined as the use of multiple tactics in a compatible manner to maintain pest populations at levels below those causing economic injury while providing protection against hazards to humans, animals, plants and the environment. IPM is thus ecologically-based pest management that makes full use of natural and cultural processes and methods, including host resistance and biological control. IPM emphasises the growth of a healthy crop with the least possible disruption of agro-ecosystems, thereby encouraging natural pest control mechanisms. Chemical pesticides are used only where and when these natural methods fail to keep pests below damaging levels".

The basis of this natural method of controlling pests is the biodiversity of the agroecological system. This is because the greater the diversity of natural enemy species, the lower the density of the pest population, and as diversity of natural enemy species decreases, pest population increases.

The key components of an EPM approach are:

- Crop Management: Selecting appropriate crops for local climate and soil conditions. Practices include:
    - Selection of pest-resistant, local, native varieties and well adapted cultivars.
    - Use of legume-based crop rotations to increase soil nitrate availability thereby improving soil fertility and favourable conditions for robust plants that better face pests and diseases.
    - Use of cover crops, such as green manure to reduce weed infestation, disease and pest

attacks. This is based in the competence between different plants species. The idea is to leave the least space possible for the so-called 'bad herbs' (brush) to be established so that they do not compete for the soil nutrients with the crops and they do not become hosts of plagues (insects) or crop diseases (bacteria and fungi).

- Integration of intercropping and agro-forestry systems.
- Use of crop spacing, intercropping and pruning to create conditions unfavourable to the pests.

- Soil Management: Maintaining soil nutrition and pH levels to provide the best possible chemical, physical, and biological soil habitat for crops. Practices include:
  - Building a healthy soil (soil relatively free of fungi, bacteria and insects, with basic nutrients (nitrogen, phosphorus, and potassium) in optimal level and with acidity or alkalinity levels (pH level) that make them available for crops) structure according to the soil requirements of the different plants (such as deep/shallow soil levels or different mineral contents).
  - Using longer crop rotations to enhance soil microbial populations and break disease, insect and weed cycles.
  - Applying organic manures to help maintain balanced pH and nutrient levels. Adding earthworm castings, colloidal minerals, and soil inoculants will supplement this. Microbes in the compost will improve water absorption and air exchange.
  - Soil nutrients can be reactivated by alleviating soil compaction.
  - Reducing soil disturbance (tillage) - undisturbed soil with sufficient supply of organic matter provides a good habitat for soil fauna.
  - Keeping soil covered with crop residue or living plants.
  - Releasing beneficial insects and providing them with a suitable habitat.
  - Managing plant density and structure so as to deter diseases.
  - Cultivating for weed control based on knowledge of the critical competition period.
  - Managing field boundaries and in-field habitats to attract beneficial insects, and trap or confuse insect pests.
- Pest Management: Using beneficial organisms that behave as parasitoids and predators. Practices include:
  - Releasing beneficial insects and providing them with a suitable habitat.
  - Managing plant density and structure so as to deter diseases.
  - Cultivating for weed control based on knowledge of the critical competition period.
  - Managing field boundaries and in-field habitats to attract beneficial insects, and trap or confuse insect pests.

- IPM strategies can exist at various levels of integration:
  - Control of a single pest on a particular crop.
  - Control of several pests on the same crop.
  - Several crops (and non-crop species) within a single production unit (farm).
  - Several farms in a region (area-wide pest management).
- These practices, if well implemented, result in systems that are:
  - Self-regulating, maintaining populations of pests within acceptable boundaries.
  - Self-sufficient, with minimal need for 'reactive' interventions.
  - Resistant to stresses such as drought, soil compaction, pest invasions.
  - Capable of recuperating from stresses.

EPM is a biotechnology belonging to the denominated 'clean' technologies which combines the life cycle of crops, insects and implicated fungi, with natural external inputs (i.e. bio-pesticides) that allows a better guarantee of good harvesting even in difficult conditions of pests and diseases that emerge with the temperature and water level changes (increase of relative atmospheric humidity and runoff) typical of climate change. Thus, it is a biotechnology for facing uncertainty caused by climate change.

EPM contributes to climate change adaptation by providing a healthy and balanced ecosystem in which the vulnerability of plants to pests and diseases is decreased. By promoting a diversified farming system, the practice of EPM builds farmers' resilience to potential risks posed by climate change, such as damage to crop yields caused by newly emerging pests and diseases.

Advantages of the Technology

With the EPM approach, farmers can avoid the costs of pesticides as well as the fuel, equipment and labour used to apply them. A 22-year trial comparing conventional and organic corn/soybean systems found that organic farming approaches for these crops use an average of 30 per cent less fossil energy. Although this can cause a slight drop in productive performance, the risk of losing an entire crop is reduced dramatically.

There are also reports that production levels have increased when there has been a reduction in the use of pesticides. This is the case when there are specific controllers for a determined pest, for example, in West Africa the introduction of the wasp has been a spectacular control of the slug of cassava, thus saving the staple food crop for millions of Africans.

Disadvantages of the Technology

There are very strong pests for which the 'biological controller' has not yet been identified (i.e. an insect that destroys it). When these pests emerge it is common for producers to turn to pesticides. EPM is not easy to implement and requires substantial knowledge and monitoring for the combined components of the system to produce success. Perhaps the biggest drawback to the EPM approach is that biological control is not a 'quick fix'. In most cases, biological controllers will take several years to successfully establish a population and begin making a significant contribution. In addition, no single biological controller works in every situation. A controller that works well in one soil type, for example, may not work at all in another soil type. In the long run, more than one type of biological controller may have to be used to achieve uniform control across a variety of different situations and land types.

Institutional and Organisational Requirements

Structures that enable farmers to organise themselves so as to jointly implement the proposed solutions are also required. Collective action can increase the successful development and implementation of EPM. Growers' cooperation can help reduce the costs of EPM implementation. In addition, better linkages between research and extension, more extension services, monitoring services and private consultants can lead to better coordination and feedback processes.

Strong efforts in the area of communication with farmers are required so that they appreciate the benefits of applying this approach. Communication should be primarily focused on showing the range of advantages of this technology in comparison with other available options (such as longer-term sustainability and no environmental damage). Public sector agencies, such as ministries of environment, should lead on these initiatives.

Barriers to Implementation

Major constraints to the development and adoption of EPM programmes fall into four categories:

1. Technical: Lack of studies and complexity of EPM.
2. Economic: Competing simplicity and apparent efficacy of chemicals; lower prices for EPM-produced goods (cosmetic damage); high cost of selective pesticides; lack of fiscal policy that favours EPM over pesticide use; high perceived risk if spraying is not carried out; failure to consider long-term advantages). A major obstacle to the implementation of this technology is that farmers generally prefer commercial pesticides because they are easier to apply and manage.
3. Institutional: Poor linkages between research and extension; lack of extension services, monitoring services, private consultants.
4. Educational: Lack of understanding of EPM by farmers/extension, lack of EPM specialists.

EPM is complex and for farmers to understand and adopt EPM strategies they frequently have to change their whole pest control philosophy. There is also a common misconception that pesticides are essential for high yields.

BIOEFFECTOR

A Bioeffector is a viable microorganism or active natural compound which directly or indirectly affects plant performance (Biofertilizer), and thus has the potential to reduce fertilizer and pesticide use in crop production.

Types

Bioeffectors have a direct or indirect effect on plant performance by influencing the functional implementation or activation of biological mechanisms, particularly those interfering with soil-plant-microbe interactions. In contrast to conventional fertilizers and pesticides, the effectiveness of bioeffectors is not based on a substantial direct input of mineral plant nutrients, either in inorganic or organic forms.

- Products in use are:
  - Microbial residues,
  - Composting and fermentation products,
  - Plant and algae extracts.
- Bioeffector-preparations (*bio-agents*) as ready-formulated products are applied:
  - With the purpose of stimulating plant growth (bio-stimulants),
  - To improve plant nutrient acquisition (bio-fertilizers),
  - To protect plants from pathogens and pests (bio-control agents),
  - Or generally to advance cropping efficiency; they can contain one or more bio-effectors along with other substances".
- Well established bioeffectors with documented positive results in the field level are:
  - Rhizobia strains for soil or seed inoculation as a prerequisite for symbiotic N_2-fixation when establishing new legume species or varieties.
  - Positive effects of mycorrhiza inoculation for soils with a (temporarily) low potential for natural root mycorrhization.
  - Sufficient mycorrhization enhances nutrient (P) and water uptake and increases resistance to pathogenic fungi.
- Further mechanisms for the positive impact of bioeffectors on plant growth have postulated, promising a high potential for resource preservation due to reduction of fertiliser and pesticide use:
  - Active nutrient mobilisation by exudation of acids and carboxylates (e.g. P-mobilisation),
  - Exudation of micro-nutrient mobilising siderophores/chelates (e.g. Fe^{3+}),
  - Reduction of trace elements from less soluble oxidised to highly soluble reduced forms (e.g. Fe^{3+} to Fe^{2+}, Mn^{4+} to Mn^{2+}),
  - Associative/non-symbiotic N_2-fixation, protective antagonism to plant pathogens,
  - Enhancement of mycorrhizal infection and growth, and stimulating hormonal effects.

BIOLOGICAL PEST CONTROL

Syrphus hoverfly larva (below) feed on aphids (above), making them natural biological control agents.

Biological control or biocontrol is a method of controlling pests such as insects, mites, weeds and plant diseases using other organisms. It relies on predation, parasitism, herbivory, or other natural mechanisms, but typically also involves an active human management role. It can be an important component of integrated pest management (IPM) programs.

There are three basic strategies for biological pest control: classical (importation), where a natural enemy of a pest is introduced in the hope of achieving control; inductive (augmentation), in which a large population of natural enemies are administered for quick pest control; and inoculative (conservation), in which measures are taken to maintain natural enemies through regular reestablishment.

A parasitoid wasp (Cotesia congregata) adult with pupal cocoons on its host, a tobacco hornworm (Manduca sexta, green background), an example of a hymenopteran biological control agent.

Natural enemies of insect pests, also known as biological control agents, include predators, parasitoids, pathogens, and competitors. Biological control agents of plant diseases are most often referred to as antagonists. Biological control agents of weeds include seed predators, herbivores and plant pathogens.

Biological control can have side-effects on biodiversity through attacks on non-target species by any of the same mechanisms, especially when a species is introduced without thorough understanding of the possible consequences.

Types of Biological Pest Control

There are three basic biological pest control strategies: importation (classical biological control), augmentation and conservation.

Importation

Rodolia cardinalis, the vedalia beetle, was imported from Australia to California in the 19th century, successfully controlling cottony cushion scale.

Importation or classical biological control involves the introduction of a pest's natural enemies to a new locale where they do not occur naturally. Early instances were often unofficial and not based on research, and some introduced species became serious pests themselves.

To be most effective at controlling a pest, a biological control agent requires a colonizing ability which allows it to keep pace with changes to the habitat in space and time. Control is greatest if the agent has temporal persistence, so that it can maintain its population even in the temporary absence of the target species, and if it is an opportunistic forager, enabling it to rapidly exploit a pest population.

Joseph Needham noted a Chinese text dating from 304 AD, Records of the Plants and Trees of the Southern Regions, by Hsi Han, which describes mandarin oranges protected by large reddish-yellow citrus ants which attack and kill insect pests of the orange trees. The citrus ant (Oecophylla smaragdina) was rediscovered in the 20th century, and since 1958 has been used in China to protect orange groves.

One of the earliest successes in the west was in controlling Icerya purchasi (cottony cushion scale) in Australia, using a predatory insect Rodolia cardinalis (the vedalia beetle). This success was repeated in California using the beetle and a parasitoidal fly, Cryptochaetum iceryae. Other successful cases include the control of Antonina graminis in Texas by Neodusmetia sangwani in the 1960s.

Damage from Hypera postica, the alfalfa weevil, a serious introduced pest of forage, was substantially reduced by the introduction of natural enemies. 20 years after their introduction the population of weevils in the alfalfa area treated for alfalfa weevil in the Northeastern United States remained 75 percent down.

The invasive species Alternanthera philoxeroides (alligator weed) was controlled in Florida (U.S.) by introducing alligator weed flea beetle.

Alligator weed was introduced to the United States from South America. It takes root in shallow water, interfering with navigation, irrigation, and flood control. The alligator weed flea beetle and two other biological controls were released in Florida, greatly reducing the amount of land covered by the plant. Another aquatic weed, the giant salvinia (Salvinia molesta) is a serious pest, covering waterways, reducing water flow and harming native species. Control with the salvinia weevil (Cyrtobagous salviniae) and the salvinia stem-borer moth (Samea multiplicalis) is effective in warm climates, and in Zimbabwe, a 99% control of the weed was obtained over a two-year period.

Small commercially reared parasitoidal wasps, Trichogramma ostriniae, provide limited and erratic control of the European corn borer (Ostrinia nubilalis), a serious pest. Careful formulations of the bacterium Bacillus thuringiensis are more effective.

The population of Levuana iridescens, the Levuana moth, a serious coconut pest in Fiji, was brought under control by a classical biological control program in the 1920s.

Augmentation

Augmentation involves the supplemental release of natural enemies that occur in a particular area, boosting the naturally occurring populations there. In inoculative release, small numbers of the control agents are released at intervals to allow them to reproduce, in the hope of setting up longer-term control, and thus keeping the pest down to a low level, constituting prevention rather than cure. In inundative release, in contrast, large numbers are released in the hope of rapidly reducing a damaging pest population, correcting a problem that has already arisen. Augmentation can be effective, but is not guaranteed to work, and depends on the precise details of the interactions between each pest and control agent.

An example of inoculative release occurs in the horticultural production of several crops in greenhouses. Periodic releases of the parasitoidal wasp, *Encarsia formosa*, are used to control greenhouse whitefly, while the predatory mite *Phytoseiulus persimilis* is used for control of the two-spotted spider mite.

The egg parasite Trichogramma is frequently released inundatively to control harmful moths. Similarly, Bacillus thuringiensis and other microbial insecticides are used in large enough quantities for a rapid effect. Recommended release rates for Trichogramma in vegetable or field crops range from 5,000 to 200,000 per acre (1 to 50 per square metre) per week according to the level of pest infestation. Similarly, nematodes that kill insects (that are entomopathogenic) are released at rates of millions and even billions per acre for control of certain soil-dwelling insect pests.

Conservation

The conservation of existing natural enemies in an environment is the third method of biological pest control. Natural enemies are already adapted to the habitat and to the target pest, and their conservation can be simple and cost-effective, as when nectar-producing crop plants are grown in the borders of rice fields. These provide nectar to support parasitoids and predators of planthopper pests and have been demonstrated to be so effective (reducing pest densities by 10- or even 100-fold) that farmers sprayed 70% less insecticides and enjoyed yields boosted by 5%. Predators of aphids were similarly found to be present in tussock grasses by field boundary hedges in England, but they spread too slowly to reach the centres of fields. Control was improved by planting a metre-wide strip of tussock grasses in field centres, enabling aphid predators to overwinter there.

An inverted flowerpot filled with straw to attract earwigs.

Cropping systems can be modified to favor natural enemies, a practice sometimes referred to as habitat manipulation. Providing a suitable habitat, such as a shelterbelt, hedgerow, or beetle bank where beneficial insects such as parasitoidal wasps can live and reproduce, can help ensure the survival of populations of natural enemies. Things as simple as leaving a layer of fallen leaves or mulch in place provides a suitable food source for worms and provides a shelter for insects, in turn being a food source for such beneficial mammals as hedgehogs and shrews. Compost piles and stacks of wood can provide shelter for invertebrates and small mammals. Long grass and ponds support amphibians. Not removing dead annuals and non-hardy plants in the autumn allows insects to make use of their hollow stems during winter. In California, prune trees are sometimes planted in grape vineyards to provide an improved overwintering habitat or refuge for a key grape pest parasitoid. The providing of artificial shelters in the form of wooden caskets, boxes or flowerpots is also sometimes undertaken, particularly in gardens, to make a cropped area more attractive to natural enemies. For example, earwigs are natural predators which can be encouraged in gardens by hanging upside-down flowerpots filled with straw or wood wool. Green lacewings can be encouraged by using plastic bottles with an open bottom and a roll of cardboard inside. Birdhouses enable insectivorous birds to nest; the most useful birds can be attracted by choosing an opening just large enough for the desired species.

In cotton production, the replacement of broad-spectrum insecticides with selective control measures such as Bt cotton can create a more favorable environment for natural enemies of cotton pests due to reduced insecticide exposure risk. Such predators or parasitoids can control pests not affected by the Bt protein. Reduced prey quality and abundance associated increased control from Bt cotton can also indirectly decrease natural enemy populations in some cases, but the percentage of pests eaten or parasitized in Bt and non-Bt cotton are often similar.

Biological Control Agents

Predators

Predatory lacewings are available from biocontrol dealers.

Predators are mainly free-living species that directly consume a large number of prey during their whole lifetime. Given that many major crop pests are insects, many of the predators used in biological control are insectivorous species. Lady beetles, and in particular their larvae which are active between May and July in the northern hemisphere, are voracious predators of aphids, and also consume mites, scale insects and small caterpillars. The spotted lady beetle (Coleomegilla maculata) is also able to feed on the eggs and larvae of the Colorado potato beetle (Leptinotarsa decemlineata).

The larvae of many hoverfly species principally feed upon aphids, one larva devouring up to 400 in its lifetime. Their effectiveness in commercial crops has not been studied.

Predatory *Polistes* wasp searching for bollworms or other caterpillars on a plant.

Several species of entomopathogenic nematode are important predators of insect and other invertebrate pests. Phasmarhabditis hermaphrodita is a microscopic nematode that kills slugs. Its complex life cycle includes a free-living, infective stage in the soil where it becomes associated with a pathogenic bacteria such as Moraxella osloensis. The nematode enters the slug through the posterior mantle region, thereafter feeding and reproducing inside, but it is the bacteria that kill the slug. The nematode is available commercially in Europe and is applied by watering onto moist soil.

Species used to control spider mites include the predatory mites Phytoseiulus persimilis, Neoseilus californicus, and Amblyseius cucumeris, the predatory midge Feltiella acarisuga, and a ladybird Stethorus punctillum. The bug Orius insidiosus has been successfully used against the two-spotted spider mite and the western flower thrips (Frankliniella occidentalis).

Predators including Cactoblastis cactorum (mentioned above) can also be used to destroy invasive plant species. As another example, the poison hemlock moth (Agonopterix alstroemeriana) can be used to control poison hemlock (Conium maculatum). During its larval stage, the moth strictly consumes its host plant, poison hemlock, and can exist at hundreds of larvae per individual host plant, destroying large swathes of the hemlock.

The parasitoid wasp Aleiodes indiscretus parasitizing a gypsy moth caterpillar, a serious pest of forestry.

For rodent pests, cats are effective biological control when used in conjunction with reduction of "harborage"/hiding locations. While cats are effective at preventing rodent "population explosions", they are not effective for eliminating pre-existing severe infestations. Barn owls are also sometimes used as biological rodent control. Although there are no quantitative studies of the effectiveness of barn owls for this purpose, they are known rodent predators that can be used in addition to or instead of cats; they can be encouraged into an area with nest boxes.

Parasitoids

Parasitoids lay their eggs on or in the body of an insect host, which is then used as a food for developing larvae. The host is ultimately killed. Most insect parasitoids are wasps or flies, and many have a very narrow host range. The most important groups are the ichneumonid wasps, which mainly use caterpillars as hosts; braconid wasps, which attack caterpillars and a wide range of other insects including aphids; chalcid wasps, which parasitize eggs and larvae of many insect species; and tachinid flies, which parasitize a wide range of insects including caterpillars, beetle adults and larvae, and true bugs. Parasitoids are most effective at reducing pest populations when their host organisms have limited refuges to hide from them.

Parasitoids lay their eggs on or in the body of an insect host.

Parasitoids are among the most widely used biological control agents. Commercially, there are two types of rearing systems: short-term daily output with high production of parasitoids per day, and long-term, low daily output systems. In most instances, production will need to be matched with the appropriate release dates when susceptible host species at a suitable phase of development will be available. Larger production facilities produce on a yearlong basis, whereas some facilities produce only seasonally. Rearing facilities are usually a significant distance from where the agents are to be used in the field, and transporting the parasitoids from the point of production to the point of use can pose problems. Shipping conditions can be too hot, and even vibrations from planes or trucks can adversely affect parasitoids.

Encarsia formosa is a small predatory chalcid wasp which is a parasitoid of whitefly, a sap-feeding insect which can cause wilting and black sooty moulds in glasshouse vegetable and ornamental crops. It is most effective when dealing with low level infestations, giving protection over a long period of time. The wasp lays its eggs in young whitefly 'scales', turning them black as the parasite larvae pupate. Gonatocerus ashmeadi (Hymenoptera: Mymaridae) has been introduced to control the glassy-winged sharpshooter Homalodisca vitripennis (Hemiptera: Cicadellidae) in French Polynesia and has successfully controlled ~95% of the pest density.

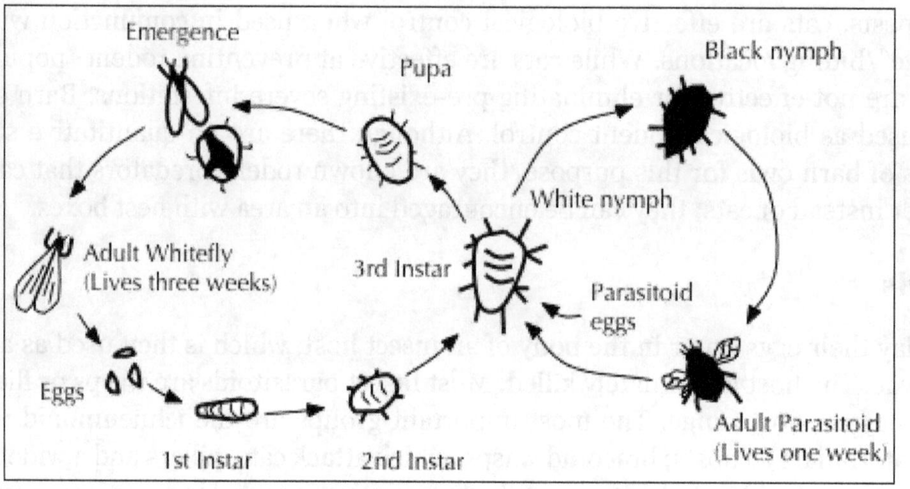

Life cycles of greenhouse whitefly and its parasitoid wasp *Encarsia formosa.*

The eastern spruce budworm is an example of a destructive insect in fir and spruce forests. Birds are a natural form of biological control, but the Trichogramma minutum, a species of parasitic wasp, has been investigated as an alternative to more controversial chemical controls.

There are a number of recent studies pursuing sustainable methods for controlling urban cockroaches using parasitic wasps. Since most cockroaches remain in the sewer system and sheltered areas which are inaccessible to insecticides, employing active-hunter wasps is a strategy to try and reduce their populations.

Pathogens

Pathogenic micro-organisms include bacteria, fungi, and viruses. They kill or debilitate their host and are relatively host-specific. Various microbial insect diseases occur naturally, but may also be used as biological pesticides. When naturally occurring, these outbreaks are density-dependent in that they generally only occur as insect populations become denser.

Bacteria

Bacteria used for biological control infect insects via their digestive tracts, so they offer only limited options for controlling insects with sucking mouth parts such as aphids and scale insects. Bacillus thuringiensis, a soil-dwelling bacterium, is the most widely applied species of bacteria used for biological control, with at least four sub-species used against Lepidopteran (moth, butterfly), Coleopteran (beetle) and Dipteran (true fly) insect pests. The bacterium is available to organic farmers in sachets of dried spores which are mixed with water and sprayed onto vulnerable plants such as brassicas and fruit trees. Genes from B. thuringiensis have also been incorporated into transgenic crops, making the plants express some of the bacterium's toxins, which are proteins. These confer resistance to insect pests and thus reduce the necessity for pesticide use. If pests develop resistance to the toxins in these crops, B. thuringiensis will become useless in organic farming also. The bacterium Paenibacillus popilliae which causes milky spore disease has been found useful in the control of Japanese beetle, killing the larvae. It is very specific to its host species and is harmless to vertebrates and other invertebrates.

Fungi

Green peach aphid, a pest in its own right and a vector of plant viruses, killed by the fungus Pandora neoaphidis.

Entomopathogenic fungi, which cause disease in insects, include at least 14 species that attack aphids. Beauveria bassiana is mass-produced and used to manage a wide variety of insect pests including whiteflies, thrips, aphids and weevils. Lecanicillium spp. are deployed against white flies, thrips and aphids. Metarhizium spp. are used against pests including beetles, locusts and other grasshoppers, Hemiptera, and spider mites. Paecilomyces fumosoroseus is effective against white flies, thrips and aphids; Purpureocillium lilacinus is used against root-knot nematodes, and 89 Trichoderma species against certain plant pathogens. Trichoderma viride has been used against Dutch elm disease, and has shown some effect in suppressing silver leaf, a disease of stone fruits caused by the pathogenic fungus Chondrostereum purpureum.

The fungi Cordyceps and Metacordyceps are deployed against a wide spectrum of arthropods. Entomophaga is effective against pests such as the green peach aphid.

Several members of Chytridiomycota and Blastocladiomycota have been explored as agents of biological control. From Chytridiomycota, Synchytrium solstitiale is being considered as a control agent of the yellow star thistle (Centaurea solstitialis) in the United States.

Viruses

Baculoviruses are specific to individual insect host species and have been shown to be useful in biological pest control. For example, the Lymantria dispar multicapsid nuclear polyhedrosis virus has been used to spray large areas of forest in North America where larvae of the gypsy moth are causing serious defoliation. The moth larvae are killed by the virus they have eaten and die, the disintegrating cadavers leaving virus particles on the foliage to infect other larvae.

A mammalian virus, the rabbit haemorrhagic disease virus was introduced to Australia to attempt to control the European rabbit populations there. It escaped from quarantine and spread across the country, killing large numbers of rabbits. Very young animals survived, passing immunity to their offspring in due course and eventually producing a virus-resistant population. Introduction into New Zealand in the 1990s was similarly successful at first, but a decade later, immunity had developed and populations had returned to pre-RHD levels.

Oomycota

Lagenidium giganteum is a water-borne mold that parasitizes the larval stage of mosquitoes. When applied to water, the motile spores avoid unsuitable host species and search out suitable mosquito larval hosts. This mold has the advantages of a dormant phase, resistant to desiccation, with slow-release characteristics over several years. Unfortunately, it is susceptible to many chemicals used in mosquito abatement programmes.

Competitors

The legume vine Mucuna pruriens is used in the countries of Benin and Vietnam as a biological control for problematic Imperata cylindrica grass: the vine is extremely vigorous and suppresses neighbouring plants by out-competing them for space and light. Mucuna pruriens is said not to be invasive outside its cultivated area. Desmodium uncinatum can be used in push-pull farming to stop the parasitic plant, witchweed (Striga).

The Australian bush fly, *Musca vetustissima*, is a major nuisance pest in Australia, but native decomposers found in Australia are not adapted to feeding on cow dung, which is where bush flies breed. Therefore, the Australian Dung Beetle Project, led by George Bornemissza of the Commonwealth Scientific and Industrial Research Organisation, released forty-nine species of dung beetle, to reduce the amount of dung and therefore also the potential breeding sites of the fly.

Combined use of Parasitoids and Pathogens

In cases of massive and severe infection of invasive pests, techniques of pest control are often used in combination. An example is the emerald ash borer, Agrilus planipennis, an invasive beetle from China, which has destroyed tens of millions of ash trees in its introduced range in North America. As part of the campaign against it, from 2003 American scientists and the Chinese Academy of Forestry searched for its natural enemies in the wild, leading to the discovery of several parasitoid wasps, namely Tetrastichus planipennisi, a gregarious larval endoparasitoid, Oobius agrili, a solitary, parthenogenic egg parasitoid, and Spathius agrili, a gregarious larval ectoparasitoid. These have been introduced and released into the United States of America as a possible biological control of the emerald ash borer. Initial results for Tetrastichus planipennisi have shown promise, and it is now being released along with Beauveria bassiana, a fungal pathogen with known insecticidal properties.

Difficulties

Many of the most important pests are exotic, invasive species that severely impact agriculture, horticulture, forestry and urban environments. They tend to arrive without their co-evolved parasites, pathogens and predators, and by escaping from these, populations may soar. Importing the natural enemies of these pests may seem a logical move but this may have unintended consequences; regulations may be ineffective and there may be unanticipated effects on biodiversity, and the adoption of the techniques may prove challenging because of a lack of knowledge among farmers and growers.

Side Effects

Biological control can affect biodiversity through predation, parasitism, pathogenicity, competition, or other attacks on non-target species. An introduced control does not always target only the intended pest species; it can also target native species. In Hawaii during the 1940s parasitic wasps were introduced to control a lepidopteran pest and the wasps are still found there today. This may have a negative impact on the native ecosystem; however, host range and impacts need to be studied before declaring their impact on the environment.

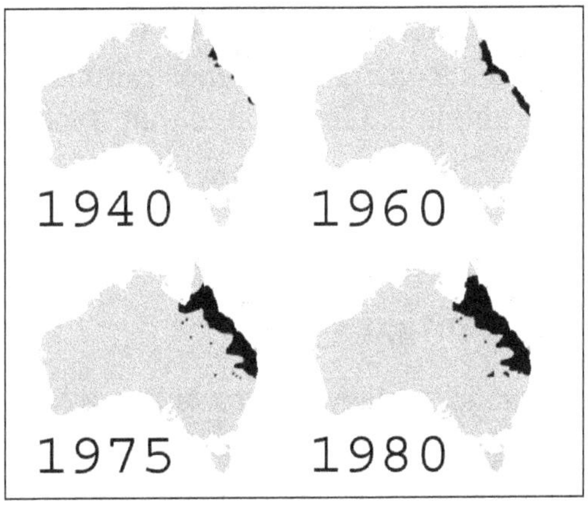

Cane toad spread from 1940 to 1980: it was ineffective as a control agent. Its distribution has continued to widen since 1980.

Vertebrate animals tend to be generalist feeders, and seldom make good biological control agents; many of the classic cases of "biocontrol gone awry" involve vertebrates. For example, the cane toad (Rhinella marina) was intentionally introduced to Australia to control the greyback cane beetle (Dermolepida albohirtum), and other pests of sugar cane. 102 toads were obtained from Hawaii and bred in captivity to increase their numbers until they were released into the sugar cane fields of the tropic north in 1935. It was later discovered that the toads could not jump very high and so were unable to eat the cane beetles which stayed on the upper stalks of the cane plants. However, the toad thrived by feeding on other insects and soon spread very rapidly; it took over native amphibian habitat and brought foreign disease to native toads and frogs, dramatically reducing their populations. Also, when it is threatened or handled, the cane toad releases poison from parotoid glands on its shoulders; native Australian species such as goannas, tiger snakes, dingos and northern quolls that attempted to eat the toad were harmed or killed. However, there has been some recent evidence that native predators are adapting, both physiologically and through changing their behaviour, so in the long run, their populations may recover.

Rhinocyllus conicus, a seed-feeding weevil, was introduced to North America to control exotic musk thistle (Carduus nutans) and Canadian thistle (Cirsium arvense). However, the weevil also attacks native thistles, harming such species as the endemic Platte thistle (Cirsium neomexicanum) by selecting larger plants (which reduced the gene pool), reducing seed production and ultimately threatening the species' survival. Similarly, the weevil Larinus planus was also used to try to control the Canadian thistle, but it damaged other thistles as well. This included one species classified as threatened.

The small Asian mongoose (Herpestus javanicus) was introduced to Hawaii in order to control the rat population. However, the mongoose was diurnal, and the rats emerged at night; the mongoose therefore preyed on the endemic birds of Hawaii, especially their eggs, more often than it ate the rats, and now both rats and mongooses threaten the birds. This introduction was undertaken without understanding the consequences of such an action. No regulations existed at the time, and more careful evaluation should prevent such releases now.

The sturdy and prolific eastern mosquitofish (Gambusia holbrooki) is a native of the southeastern United States and was introduced around the world in the 1930s and '40s to feed on mosquito larvae and thus combat malaria. However, it has thrived at the expense of local species, causing a decline of endemic fish and frogs through competition for food resources, as well as through eating their eggs and larvae. In Australia, control of the mosquitofish is the subject of discussion; in 1989 researchers A. H. Arthington and L. L. Lloyd stated that "biological population control is well beyond present capabilities".

Advantages and Disadvantages of Biological Weed Control

Biological Weed Control Advantages

Biological Pest control is a very specific strategy, whatever the predator is introduced will only control the population of the pest they are meant to target, making it a green alternative to the chemical or the mechanical control methods, whereas the weed killing chemicals can destroy fruit-bearing plants, The biological control allows the fruit to be left uninterrupted while the weeds are destroyed.

Biological pest control does not have adverse effect on the human health or the environment. It is self-sustaining, It can be cost effective. When the cost of testing and introducing control agents has been met, the on-going costs are small. There is no need to find and identify every individual weed to be treated. An effective agent will search out all suitable plants of the weed.

Development of the host resistance is not a problem. The biological pest control is compatible with most other control techniques (except sometimes the use of the insecticides and the herbicides), The biological control reduces the competitiveness and reproductive capacity of the weed, making it more manageable.

The biological control method of pest management does not use the chemicals. It uses many organisms that are either the predators or the parasites to the pest. The pest is the organism that causes damage to the people and their crops. The biological control should be implemented whenever possible because it does not pollute the environment.

The great benefit of this method is its selectivity. There is a restricted danger of damage to non target plant species. Biological control does not create new problems, like conventional pesticides. Selectivity is the most important factor regarding the balance of the agricultural ecosystems because a great damage to non target species can lead to the restriction of natural enemies' populations.

The biological control agent (BCA) can be deployed in the agricultural ecosystem, so as not to damage non target pests, depends on appropriate host specificity tests which determine the potential host range.

The ability to self-perpetuate is an interesting advantage of the biological control method, BCAs will increase in number and spread, Because BCAs are self-propagating and dispersing, pest control is self-perpetuating too, This is quite important regarding the economic feasibility of the biological control.

Another benefit of the biological control method is the environmental safety of BCAs. The pest is unable (or very slow) to develop the resistance. The biological control can be cost effective, Its effectiveness is based on the self-perpetuation and self-propagation. So, if we establish a control agent in a specific area, it will reduce the target pest in an acceptable threshold for quite long time.

The financial benefit of the biological control is greatest in the cases when there is no other option, The biological control is very effective in the inaccessible areas. The cost efficiency of this method is that the yield benefit of the biological control is less than yield achieved by the agrochemicals, but the primary cost of BCA is lower than the chemical pesticides.

The biological control is the attractive alternative to the agrochemicals, the use of environmentally friendly alternatives to the chemical pesticides are absolutely required in agriculture. No chemicals are used, so, there's less pollution, disruption of the food chain and risk to the people eating the food that has been sprayed.

BCAs are more susceptible to the environmental conditions than the chemical control. This consequently causes the fluctuations to the pest populations. It reflects to the product quality, to the crop yield and of course to the price of the product on market, if the annual harvest of the crop is not stable, it will affect the grower's income stability.

The biological control method is environmentally safe. This method offers less risk of the residues in food chain, biological control is most suited for exotic pest that are not closely related to indigenous beneficial species. Biological control is less expensive than the chemical control.

Disadvantages of Biological Weed Control

Biological control is a slow process. It takes a lot of time and patience for the biological agents to work their magic on the pest population, whereas the other methods like the pesticides work offer immediate results. The upside to this is the long-term effect biological control provides.

It can be fickle, you can't control whatever natural enemy, you set loose in an ecosystem. While it is supposed to manage one pest, your predator will switch to a different target, they may decide eating your crops instead of the insects infesting them, but when you introduce a new species to the environment, there is a risk of disrupting the natural food chain.

Although it is cheap in the long run, the process of setting up the biological control system is a costly endeavor. A lot of planning money goes into developing the successful system. The predator you introduce may not eat the pest. It could eat useful species. The predators population may increase and get out of control. The predator may not stay in the area where it is needed.

The biological control has slow action. It lacks the immediacy of chemical control. So, During the period required until the natural enemies control the pest population, the pests may be present in the intolerable populations. The agent may become a pest itself. Frequent input is needed to maintain population balance and it needs to be large scale.

The pesticides can not be used because they will damage the biological control system. The need for a difficult transition period from insecticides to the biological control may make the introduction of the biological control unattractive for a grower. It is expensive to research and It needs a high level of skills and initial set up costs.

The biological control is unpredictable. Its unpredictability lies mostly in the fact that the natural enemies are dependent on the environmental conditions. The deployment of BCAs in a new environment requires a lot of research to succeed the desirable results because of climatic constraints.

It is difficult and sometimes expensive to develop biological control in field because it requires high qualified scientific staff. There is relatively less investment in the biological control research in compare with the chemical pesticides. The variation and changes in behaviour of natural enemies that can be caused by rearing conditions are manifold. This variation leads to inconsistent results in the biological control.

The potential agents are expensive to test for specificity. Host specificity testing may take many years to complete because of the need for thoroughness. The biological control operates over large areas, so, it can not be limited to the individual properties or paddock. It does not eradicate the pest organism completely, because if the control agent reduces the pest population too far, it destroys its own food source.

BIOPESTICIDE

Biopesticides, a contraction of 'biological pesticides', include several types of pest management intervention: through predatory, parasitic, or chemical relationships. The term has been associated historically with biological control – and by implication – the manipulation of living organisms. Regulatory positions can be influenced by public perceptions, thus:

- In the EU, biopesticides have been defined as "a form of pesticide based on micro-organisms or natural products".
- The US EPA states that they "include naturally occurring substances that control pests (biochemical pesticides), microorganisms that control pests (microbial pesticides), and pesticidal substances produced by plants containing added genetic material (plant-incorporated protectants) or PIPs".

They are obtained from organisms including plants, bacteria and other microbes, fungi, nematodes, etc. They are often important components of integrated pest management (IPM) programmes, and have received much practical attention as substitutes to synthetic chemical plant protection products (PPPs).

Types

Biopesticides can be classified into these classes:

- Microbial pesticides which consist of bacteria, entomopathogenic fungi or viruses (and sometimes includes the metabolites that bacteria or fungi produce). Entomopathogenic nematodes are also often classed as microbial pesticides, even though they are multi-cellular.
- Bio-derived chemicals: Four groups are in commercial use - pyrethrum, rotenone, neem oil, and various essential oils are naturally occurring substances that control (or monitor in the case of pheromones) pests and microbial diseases.
- Plant-incorporated protectants (PIPs) have genetic material from other species incorporated into their genetic material (*i.e.* GM crops). Their use is controversial, especially in many European countries.
- RNAi pesticides, some of which are topical and some of which are absorbed by the crop.

Biopesticides have usually no known function in photosynthesis, growth or other basic aspects of plant physiology. Instead, they are active against biological pests. Many chemical compounds have been identified that are produced by plants to protect them from pests so they are called antifeedants. These materials are biodegradable and renewable alternatives, which can be economical for practical use. Organic farming systems embraces this approach to pest control.

RNA

RNA interference is under study for possible use as a spray-on insecticide by multiple companies, including Monsanto, Syngenta, and Bayer. Such sprays do not modify the genome of the target plant. The RNA could be modified to maintain its effectiveness as target species evolve tolerance to the original. RNA is a relatively fragile molecule that generally degrades within days or weeks of application. Monsanto estimated costs to be on the order of $5/acre.

RNAi has been used to target weeds that tolerate Monsanto's Roundup herbicide. RNAi mixed with a silicone surfactant that let the RNA molecules enter air-exchange holes in the plant's surface that disrupted the gene for tolerance, affecting it long enough to let the herbicide work. This strategy would allow the continued use of glyphosate-based herbicides, but would not per se assist a herbicide rotation strategy that relied on alternating Roundup with others.

They can be made with enough precision to kill some insect species, while not harming others. Monsanto is also developing an RNA spray to kill potato beetles. One challenge is to make it linger on the plant for a week, even if it's raining. The Potato beetle has become resistant to more than 60 conventional insecticides.

Monsanto lobbied the U.S. EPA to exempt RNAi pesticide products from any specific regulations (beyond those that apply to all pesticides) and be exempted from rodent toxicity, allergenicity and residual environmental testing. In 2014 an EPA advisory group found little evidence of a risk to people from eating RNA.

However, in 2012, the Australian Safe Food Foundation posited that the RNA trigger designed to change the starch content of wheat might interfere with the gene for a human liver enzyme. Supporters countered that RNA does not appear to make it past human saliva or stomach acids. The US National Honey Bee Advisory Board told EPA that using RNAi would put natural systems at "the epitome of risk". The beekeepers cautioned that pollinators could be hurt by unintended effects and that the genomes of many insects are still unknown. Other unassessed risks include ecological (given the need for sustained presence for herbicide and other applications) and the possible for RNA drift across species boundaries.

Monsanto has invested in multiple companies for their RNA expertise, including Beeologics (for RNA that kills a parasitic mite that infests hives and for manufacturing technology) and Preceres (nanoparticle lipidoid coatings) and licensed technology from Alnylam and Tekmira. In 2012 Syngenta acquired Devgen, a European RNA partner. Startup Forrest Innovations is investigating RNAi as a solution to citrus greening disease that in 2014 caused 22 percent of oranges in Florida to fall off the trees.

Examples:

Bacillus thuringiensis, a bacterial disease of Lepidoptera, Coleoptera and Diptera, is a well-known insecticide example. The toxin from B. thuringiensis (Bt toxin) has been incorporated directly into plants through the use of genetic engineering. The use of Bt Toxin is particularly controversial. Its manufacturers claim it has little effect on other organisms, and is more environmentally friendly than synthetic pesticides.

Other microbial control agents include products based on:

- Entomopathogenic fungi (e.g. Beauveria bassiana, Isaria fumosorosea, Lecanicillium and Metarhizium spp.).
- Plant disease control agents: It include Trichoderma spp. and Ampelomyces quisqualis (a hyper-parasite of grape powdery mildew); Bacillus subtilis is also used to control plant pathogens.
- Beneficial nematodes attacking insect (e.g. Steinernema feltiae) or slug (e.g. Phasmarhabditis hermaphrodita) pests.
- Entomopathogenic viruses (e.g. Cydia pomonella granulovirus).
- Weeds and rodents have also been controlled with microbial agents.

Various naturally occurring materials, including fungal and plant extracts, have been described as biopesticides. Products in this category include:

- Insect pheromones and other semiochemicals.
- Fermentation products such as Spinosad (a macro-cyclic lactone).
- Chitosan: A plant in the presence of this product will naturally induce systemic resistance (ISR) to allow the plant to defend itself against disease, pathogens and pests.
- Biopesticides may include natural plant-derived products, which include alkaloids, terpenoids, phenolics and other secondary chemicals. Certain vegetable oils such as canola oil

are known to have pesticidal properties. Products based on extracts of plants such as garlic have now been registered in the EU and elsewhere.

Applications

Biopesticides are biological or biologically-derived agents, that are usually applied in a manner similar to chemical pesticides, but achieve pest management in an environmentally friendly way. With all pest management products, but especially microbial agents, effective control requires appropriate formulation and application.

Biopesticides for use against crop diseases have already established themselves on a variety of crops. For example, biopesticides already play an important role in controlling downy mildew diseases. Their benefits include: a 0-Day Pre-Harvest Interval, the ability to use under moderate to severe disease pressure, and the ability to use as a tank mix or in a rotational program with other registered fungicides. Because some market studies estimate that as much as 20% of global fungicide sales are directed at downy mildew diseases, the integration of biofungicides into grape production has substantial benefits in terms of extending the useful life of other fungicides, especially those in the reduced-risk category.

A major growth area for biopesticides is in the area of seed treatments and soil amendments. Fungicidal and biofungicidal seed treatments are used to control soil borne fungal pathogens that cause seed rots, damping-off, root rot and seedling blights. They can also be used to control internal seed–borne fungal pathogens as well as fungal pathogens that are on the surface of the seed. Many biofungicidal products also show capacities to stimulate plant host defence and other physiological processes that can make treated crops more resistant to a variety of biotic and abiotic stresses.

Disadvantages

- High specificity: Which may require an exact identification of the pest/pathogen and the use of multiple products to be used; although this can also be an advantage in that the biopesticide is less likely to harm species other than the target.
- Often slow speed of action (thus making them unsuitable if a pest outbreak is an immediate threat to a crop).
- Often variable efficacy due to the influences of various biotic and abiotic factors (since some biopesticides are living organisms, which bring about pest/pathogen control by multiplying within or nearby the target pest/pathogen).
- Living organisms evolve and increase their resistance to biological, chemical, physical or any other form of control. If the target population is not exterminated or rendered incapable of reproduction, the surviving population can acquire a tolerance of whatever pressures are brought to bear, resulting in an evolutionary arms race.
- Unintended consequences: Studies have found broad spectrum biopesticides have lethal and nonlethal risks for non-target native pollinators such as *Melipona quadrifasciata* in Brazil.

Bioherbicide

Bioherbicides are biologically based control agents useful for biological weed control. Hence, bioherbicides have been identified as a significant biological control strategy. Bioherbicides have many advantages such as clearly defined target weeds, no side effect on beneficial plants or human health, a lack of pesticide residue build-up in the environment, and effectiveness for control of some herbicide-resistant weed biotypes. More importantly, it has been demonstrated that mixtures of some bioherbicides and synthetic herbicides can be more effective.

Development of alternative weed control methods is needed to help decrease reliance on herbicide use. Biological weed control is an alternative option for weed problems, particularly in agriculture and forestry. It is based on the use of natural enemies, particularly insects and pathogens to control weeds, as a sustainable, low cost and more environmentally acceptable method of weed control. One of the approaches to biological weed control using pathogens, mainly fungi, is inundative, bioherbicide approach.

Bioherbicides are phytopathogenic microorganisms or microbial phytotoxins useful for biological weed control applied in similar ways to conventional herbicides. The active ingredient in a bioherbicide is, however, a living microorganism. Most commonly the organism is a fungus; hence the term mycoherbicide is often used in these cases. Although the use of fungi and bacteria as inundative biological control agents (bioherbicides) has been recognized as a significant technological weed control alternative, it can be argued that it serves a more important role as a complementary component in successful integrated management strategies, and not as a replacement for chemical herbicides and other weed management tactics. Actually, in many situations, bioherbicides can be used as the sole option for the management of one or two target weeds, i.e. as a minor supplement to conventional chemical herbicides.

Bioherbicides offer many advantages in comparison with synthetic herbicides. They include:

- A high degree of specificity of target weed;
- No effect on non-target and beneficial plants or man;
- Absence of residue build-up in the environment;
- Effectiveness for managing herbicide-resistant (HR) weed populations.

Except numerous advantages of bioherbicides, some circumstances have been noted to restrain the progress of bioherbicides into profitable outputs. These include:

- Biological restrictions (host changeability, host scope resistance mechanisms and interaction with other microorganisms that affect efficacy);
- Environment restrictions (epidemiology of bioherbicides reliant on optimal environmental conditions);
- Technical restrictions (wholesale production and formulation development of reliable and effective bioherbicide);
- Commercial restrictions (market capacity, patent protection, confidence and adjustment).

Classical Approach

The biocontrol approach using an imported pathogen to control a native or naturalized weed with minimal manipulations has been termed the inoculative or classical biocontrol method. The classical approach is directed mainly towards the control of exotic weeds, which have spread in the introduced area in the absence of natural enemies. Control is achieved by the importation and release of highly host-specific pathogens virulent to the target weed in its native region. These agents feed on the weed, reproduce and gradually suppress the weed as their population grows.

A highly successful biological control programme was implemented in Hawaii in the 1970s when a white smut fungus (*Entyloma ageratinae* sp. nov.) was introduced from Jamaica to control the exotic weed mistflower (*Ageratina riparia*), which was invading Hawaiian indigenous forest. The effect was rapid, with 95 per cent control after 3–4 years. Western Australian golden wattle (*Acacia saligna*) is regarded as the most important invasive weed that threatens the Cape Fynbos Floristic Region of South Africa, a unique ecosystem. In about 8 years, the introduction of the rust disease (*Uromycladium tepperianum* Sacc.) from Australia had become widespread in the province and tree density was decreased by 90–95%.

Another widely acclaimed example of biological control success is the use of a rust fungus (Puccinia chondrillina) to control rush skeleton (Chondrilla juncea L.) in Australia. The rust fungus was introduced from the Mediterranean. Puccinia chondrillina was also introduced into the western United States to control a Chondrilla juncea L. biotype. However, unlike in Australia, the rust was only partially successful. An example of successful classical biological control is that of rust fungus (Puccinia carduorum), imported from Turkey and released into the northeastern United States in 1987 to control musk thistle (Carduus thoermeri). The rust fungus has spread widely from its original introduction to the western states of Wyoming and California. Baudoin et al. found that Puccinia carduorum reduces musk thistle density by rushing agedness of rust-infected musk thistle and diminishes seed production by 20– 57%. Rust fungus (Puccinia jaceae var. solstitialis), imported from Bulgaria and Turkey, was released in California, United States, in 2003 for biological control of yellow starthistle (Centaurea solstitialis L.). The host range tests on this pathogen were extensive.

Trujillo et al. have introduced septoria leaf spot (Septoria passiflorae Syd.) for the biological control of the exotic weed banana poka (Passiflora tripartita (Juss.) Poir var. tripartita), at different forest sites in Hawaii, which resulted in over 50 per cent biomass reduction of the weed 3 years after inoculation.

Klamathweed beetle (Chrysolina quadrigemina Suffrian), introduced from Australia, proved especially effective for common St. Johnswort (Hypericum perforatum L.) weed control on California rangeland. Populations of the beetles quickly grew and spread. After 5 years, millions were collected from original release sites for redistribution throughout the Pacific Northwest. Ten years after the first release, H. perforatum populations in California were reduced to less than 1% of their original size. Another example of successful classical biological control is the introduction of the black dot spurge beetle (Aphthona nigriscutis Foudras) from Hungary as a biological control agent for leafy spurge (Euphorbia esula L.). Release of these insects has resulted in a 99 per cent reduction in spurge stand density in one area and a corresponding 30-fold increase in grass biomass in pasture and rangeland after 4 years.

Bioherbicide Approach

Opposite to classical (inoculative) approach, the bioherbicide (inundative) approach uses indigenous plant pathogens that are isolated from weeds and are cultured to produce large quantity of infective material. These are utilized at amounts that will provoke tremendous levels of infection, leading to elimination of the target weed before economic damages happen. A development of this strategy involves application of weed pathogens in a manner similar to herbicide applications. Bioherbicide inoculum is susceptible to unfavourable environmental conditions after spraying, and viability needs to be maintained for as long as is necessary to achieve infection following application. Once in the field, the inundative application of inoculum is timed to coincide with the most favourable environmental conditions and susceptible growth stage of the weed, so that a disease epidemic occurs and the weed population is suppressed. Once the weed problem has been removed, natural constraints ensure that the pathogen population returns to a low level once again. The bioherbicide (inundative) approach has been successfully implemented for a number of important agricultural, invasive and exotic weeds.

Utilization of plant pathogens for weed control was first reported in the early 1900s, but the concept of using bioherbicides to control weeds attracted wide interest among weed scientists and plant pathologists after the Second World War. The earliest experiments simply involved fungus Fusarium oxysporum Schlecht. against prickly pear cactus (Opuntia ficus-indica (L.) Mill.) in Hawaii. In the 1950s, the Russians mass-produced the spores of Alternaria cuscutacidae Rudakov and applied them to the parasitic weed dodder (Cuscata spp.). In 1963, the Chinese mass-produced a different fungus (Colletotrichum gloeosporioides f. sp. cuscutae) for the same weed (Cuscata spp.). They called their mycoherbicide "LuBao" and an improved formulation is still in use today. Official date of bioherbicide control of weeds commenced in the late 1960s with an ambitious programme to find out a pathogen or pathogens for sorrels or docks (Rumex spp.) in the United States and blackberries (Rubus spp.) in Chile. From the 1970s there has been a considerable number of prosperous bioherbicide projects. The number of scientific articles on bioherbicide research has enlarged excessively since the early 1980s. The number of weeds aimed for control as well as the number of potential pathogen candidates studied has increased. Registered and unregistered uses of bioherbicides have also increased considerably. In addition, the numbers of US patents published for the bioherbicidal technology and bioherbicide handling have increased, perhaps anticipating an increased dependence on bioherbicides in the future.

Interaction between Bioherbicides and Synthetic Herbicides

The idea of combining bioherbicides with synthetic herbicides or adjuvants has been the issue of substantial research work. Moreover, it has been revealed that mixtures of some bioherbicides and synthetic herbicides can be synergistic, culminating from reduced weed defence reactions caused by the herbicides, consequently making the weeds more sensitive to pathogen attack. Christy et al. reported a synergy between trimethylsulfonium salt of glyphosate and Xanthomonas campestris against several weed species. Other synergistic interactions involving chemical herbicides and bioherbicides have been discovered and some were granted patents in the United States. According Scientists acifluorfen and bentazon were the most effective synergists and provided significant control in several weed/pathogen combinations: (Senna obtusifolia, formerly Cassia obtusifolia [L.]) and Alternaria cassiae Jurair and Khan; Aeschynomene virginica [L.] Britton, Sterns and Poggenb. and Colletotrichum gloesporioides; Sesbania exaltata Cory and Colletotrichum truncatum; and Desmodium tortuosum [SW.] DC. and Fusarium lateritium Nees. A sublethal dose of glyphosate (50 mmol L^{-1}) suppressed the biosynthesis of a phytoalexin derived from the shikimate pathway in Senna obtusifolia (L.) H. S. Irwin and Barneby, infected by Alternaria cassiae Jurair and Khan, reducing the resistance of the weed to fungal infection and disease development. Similarly, 12 DAT, Brunnichia ovata Shinners and Campsis radicans [L.] Seem. ex Bureau were controlled by 88 and 90%, respectively, through a synergistic interaction between the fungus Myrothecium verrucaria Ditmar: Fr. and the herbicide glyphosate. Neither glyphosate nor M. verrucaria controlled these weeds at commercially acceptable levels (≥80%). Accorrding to Boyette et al, timing of glyphosate application in relation to combined treatment with the bioherbicide M. verrucaria can improve the control of Pueraria lobata Ohwi, Brunnichia ovata Shinners and Campsis radicans [L.] Seem. ex Bureau. Heiny revealed that Phoma proboscis Heiny at 1×10^7 spores mL^{-1} mixured with reduced rates of 2,4-D plus MCPP controlled field bindweed (Convolvulus arvensis L.) more effectively than the herbicide mixture alone and as effectively as the pathogen at a 10-fold higher rate. Application of various crop oils and invert emulsions improved efficacy and performance of many bioherbicides and biocontrol fungi. For instance, treatment of fungus Myrothecium verrucaria (MV) strain originally isolated from sicklepod (Senna obtusifolia L.) mixture with the surfactant Silwet L-77 caused 100% mortality of Pueraria lobata Ohwi seedlings under greenhouse conditions, and 90–100% control of older Pueraria lobata Ohwi plants in naturally infested and experimental plots, respectively.

Bioherbicide Limitations

In spite of considerable research in bioherbicides, there are only a few commercially available products worldwide. This lack of availability is mainly due to limitations in bioherbicide development, which need to be overcome to ensure the future commercial success of bioherbicides. Limitations in bioherbicide development can be classified as either environmental (temperature and, particularly, humidity as major factors influencing the efficacy of bioherbicides), biological (mainly host variability and resistance), or technological–commercial (mass production and formulation, which often blocked bioherbicide development).

Environmental Limitations

Environmental limitations are a constraint to the effective use of many biological agents, including bioherbicides. Environmental factors influence formulation performance of bioherbicides as inoculum production is dependent on sporulation of the formulation. This process, although rapid, might continue over several weeks subsequent to applications and might encounter variable environmental conditions. In the application of bioherbicides, environmental conditions prevailing in the phyllosphere of plants are frequently hostile for biological control agents. A requirement of more than 12 h of dew period for severe infection by a pathogen has been reported for several potential bioherbicides and this may limit the efficacy of the bioherbicide in the field. Temperature generally has not been considered to be as critical as moisture for mycoherbicide, although field efficacy of Colletotrichum orbiculare in controlling Xanthium spinosum L. is reduced by high-temperature conditions after inoculation of plants. However, dew period length requirement and temperature typically interact. Low temperatures may greatly extend dew period length requirements for bioherbicides developed for use in crops, such as winter wheat.

Nutrient balance can play an important part in sporulation of fungi. Studies with *Colletotrichum truncatum* have shown how carbon concentration and carbon to nitrogen (C:N) ratio influence propagule production. Moreover, a defined amino acid composition of the N source improved the production of conidia. In addition, spore fitness in terms of germination and appressoria formation rate and subsequent disease production was influenced by C:N ratios.

Soil environment, moisture and the nutrient status of the soil can influence the physiology of target plants and, therefore, their interaction with aerial applied bioherbicides. Pre-emergence application has been considered as an alternative approach to overcome some of the environmental stresses imposed upon propagules applied onto the foliage or soil surface. Bioherbicides consisting of propagules of soil-borne pathogens, which normally infect at or below the soil surface, appear to be more protected from environmental extremes and may persist and give residual control. In this context, Jackson et al. reported for 95% control of the emerging Sesbania exaltata Rydb. ex A. W. Hill seedlings when Colletotrichum truncatum Andrus and Moore was incorporated into the soil.

There are many environmental limitations to applying bioherbicides and maintaining their efficacy in water as well. Auld and McRae for control of aquatic weeds a biocontrol agent would need to possess a high ecological capability to contend with varying conditions between surface and bottom, as well as across even small bodies of water. Oxygen concentration, temperature, light intensity and salinity are just four of the variables to contend with.

Biological Limitations

From a biological viewpoint, a good bioherbicide acts relatively quickly and has acceptable efficacy in control of weeds. Unfortunately, Charudattan many of the discovered weed pathogens may provide partial control of only one weed species, even under ideal conditions. This host particularity is related to the fundamental bio-physiology of the pathogen and to host changeability and resistance as well. In other words, within a population of weed species there will usually be a range of genetically diverse biotypes that may include some resistant biotypes, just as there may be a range of biotypes of microorganisms, for instance within fungal species, with slightly different host ranges, so that there is potential to mix and vary the biotypes of a species used as a bioherbicide. Non-target plant protection in relation to the potential use of Chondrostereum purpureum (Pers ex Fr.) Pouzar (silverleaf disease) to control black cherry (Prunus serotina Erhr.) in coniferous forests by modelling the dispersal of spores and therefore quantitatively assessing the risks to susceptible fruit trees outside the forest was noted by De Jong et al. Concerns have been raised regarding the potential for sexual or asexual gene exchange between bioherbicide strains and strains attacking distantly related crop plants.

Technological and Commercial Limitations

Several technological limitations have been identified that could prevent the widespread use of bioherbicides. Pathogenic strains, formulation method and the interaction of these two parameters significantly affect the shelf life of the formulations at room temperature. High concentrations and the alteration of formulations are needed to increase bioherbicide activity. Compatibility testing of formulation components that range from registered agricultural products to novel substances, such as sunscreens, humectants and starches, can consume a great deal of time and resources.

The most challenging aspect of formulating bioherbicides is to overcome the dew requirement that exists for several of them. Attempts to overcome this limitation have included developing various water-retaining materials; invert and vegetable oil emulsion formulations and granular pre-emergence formulations are considered as a promising approach to make pathogens less dependent on available water for initial infections to occur. In addition, appropriate formulations can also reduce the dosage of inoculum required to kill weeds, thus potentially reducing the cost of bioherbicides.

Experiments conducted with a number of potential bioherbicides have demonstrated that an invert emulsion allowed infection to occur in the absence of available water and reduced the need to apply high dosages of inoculum. Invert emulsions consist of a continuous oil phase that contains water droplets. Connick and Boyette have developed an invert emulsion formulation exhibiting lower viscosity and greater water-retention properties. Auld reported that application of low concentrations of vegetable oils with an emulsifying adjuvant enhances efficacy of Colletotrichum orbiculare in inciting disease on Xanthium spinosum L. in the absence of dew in greenhouse conditions. However, oil emulsions were not effective in the field conditions. An invert emulsion has been shown to overcome dew requirements and reduce the spore concentrations required. But, unfavourable characteristic is containing of more than 30% oil which makes these formulations expensive and very viscous, typically requiring special spraying equipment such as air-assist nozzles, and because of the high oil content it is likely to produce phytotoxic effects on non-target plants. Invert emulsions have been shown to cause phytotoxicity in some cases and to predispose a variety of plants to opportunistic pathogens as well.

From the other side, the main restriction in the application of solid (dry) forms of bioherbicide is that they must await suitable, moist conditions for fungal growth and infection. Moreover, during this waiting period the living active ingredients must survive in the field. In addition, ant theft has been a problem with some formulations.

The simplest liquid formulations of bioherbicides are water suspensions of spores often with a small amount of wetting agent. These are generally used as standards against which to compare more complex formulations. However, under ideal conditions for fungal infection, simple aqueous suspensions can be successful in the field Pathogenicity of an aqueous mycelial inoculum of Alternaria eichhorneae in a controlled environment experiment was improved with hydrophilic polymers such as gellan gum, alginates and the polyacrylamide. Although several polymers retained considerably more water after 6–8 h than the water-suspension controls, no increase in efficacy of the fungus Colletotrichum orbiculare was found. Vegetable oil emulsions that contain 10% oil and 1% of an emulsifying agent reduced dew dependence in controlled environment studies using C. orbiculare in control of Xanthium spinosum L. Unfortunately, in the field conditions, the efficacy of these formulations was variable.

A novel bioherbicide formulation uses a complex emulsion – water-in-oil-in-water (WOW) emulsion. It contains at least one lipophilic surfactant, at least one hydrophilic surfactant, oil and water.

Although used in the pharmaceutical, cosmetic and food industries, WOW emulsions do not appear to have been widely used in agricultural or horticultural technology. Although numerous improvements of liquid formulations of bioherbicides have been made, genetic manipulation of fungi offers a broad extent of opportunities to adjust formulations and to ameliorate bioherbicide characteristics.

BIOLOGICAL WEED CONTROL

Biological control of weeds is broadly defined as the use of an agent, a complex of agents, or biological processes to bring about weed suppression. All forms of macrobial and microbial organisms are considered as biological control agents. Examples of biological control agents include, but are not limited to: arthropods (insects and mites), plant pathogens (fungi, bacteria, viruses, and nematodes), fish, birds, and other animals. Biologically based weed management is a much broader category of approaches that may include gene modification, genetic processes, and gene products. Human activities intended to remove weeds directly or indirectly, such as hand-weeding and burning, deliberate uses of plant competition, allelopathy, and cultural and soil management practices that alter the biotic balance of soil are considered important adjuncts to biological control in integrated weed management systems.

Principles

The underlying principle behind biological approach to weed control is based on some research works that exotic plants become invasive because they have escaped from the insect herbivores and other natural enemies that limit their multiplication and distribution in their native regions; however some other factors may contribute to the tendency for particular plant species to become invasive. Therefore biological control involves using specific natural enemies that can diminish the development and reproduction of their prey organism and put some limitations to them. The predominant approach to classical biological weed control involves the importation, colonization, and establishment of exotic natural enemies (predators, parasites, and pathogens) to diminish and maintain exotic pest populations to densities that are economically insignificant.

Procedures

Some authors have outlined general procedures to be followed when embarking on classical biological weed control programs as follows: (i) evaluate the ecology, economic impact of the weed and potential conflicts of interest; (ii) survey the organisms that are already attacking the weed in the new habitat in order to distinguish accidentally introduced agents and so eliminate such from future evaluation; (iii) carry out literature search and other forms of survey to identify natural enemies attacking the weed in its native region; (iv) screen the possible biological control agents in the foreign country to determine host range and specificity, and to remove nonspecific agents from further consideration; (v) carry out further tests of promising candidates in quarantine after introduction to ensure host specificity and eliminate predators, parasites, and pathogens that may have been introduced with them; (vi) embark on mass rearing of host-specific agents; (vii) release the host-specific agents; (viii) carry out post-release evaluation to determine establishment and effectiveness of agents; and (ix) redistribute agents to other areas where control is required. Wapshere et al. presented a summary of steps normally followed when introducing a biological control agent in a classical biological control weed program as in table below.

Table: Steps normally followed when introducing a biological control agent in a classical/inoculative biological control weed program.

Steps	Details
1. Initiation	Data on taxonomy, biology, ecology, economics, native and introduced distributions, known natural enemies, etc., are compiled by initiating scientist or group. An extensive literature review is conducted on the proposed target weed and its relatives, plus known natural enemies. Conflicts of interest identified and resolved if possible
2. Target weed approval	Data in (step 1) submitted to appropriate State and Federal groups for comment; additional data may be required
3. Foreign exploration and domestic surveys	If project approved in (step 2), the center of evolution of the genus of the target weed (if known) and other suitable areas, are searched for natural enemies, particularly where these are eco-climatically similar to the area of introduction. At the same time, the weed should be investigated in the country of introduction for attacking enemies, related plants, etc.
4. Weed ecology and agent host specificity	Ecology of the target weed, its close relatives and its natural enemies is studied in the native area, and the most damaging and apparently selective agents are subjected to several years of host-specificity testing
5. Agent approval	A report on each agent is submitted to appropriate State and Federal bodies to obtain importation and release permits
6. Importation and quarantine clearance	Each agent is imported to the country of introduction where it is reared through at least one generation in quarantine to rid it of its parasites and diseases
7. Rearing and release	After a pure culture of the agent is obtained in (step 6), it is normally mass-reared and released in the field in cages or free at field sites
8. Evaluation and monitoring	Agent is monitored at field sites to determine establishment and degree of stress on target weed, or to determine reasons why the agent did not become established or efficacious
9. Redistribution	To aid spontaneous self-dissemination, agent is distributed to other areas in the target weed's distribution, if needed

Examples of Successful Biological Control of Weeds with Introduced Insects and Pathogens

One thousand one hundred and forty-four individuals (mostly entomologists and plant pathologists) have ever attended the International Symposia on Biological Control of Weeds (ISBCWs); and out of these, 450–550 weed biological control experts have been actively involved in research and development efforts over the last 50 years mainly from USA, Canada, Australia, South Africa and New Zealand. Mc Fadyen biological approach to weed control has a long history and a good success rate of 94. Frequently cited examples of successful approach to biological weed control are the prickly pear cacti (Opuntia; spp.) in Australia, eradicated by an imported moth (Cactoblastis cactorum) and rangeland in California, Oregon, Washington, and British Columbia controlled by St. John's wort Hypericum perforatum (millepertuis perforé). Mc Fadyen reported that, a list of 41 weds which have successfully been controlled using introduced insects and pathogens and another three weeds also controlled by introduced fungi applied as mycoherbicides. He stated that many of these successes have been repeated in other countries and continents. Julien presented a list of both successful and failed cases of biological weed control; this included the introduction of 225 organisms against 111 weed species, and 178 insects and 6 mites. 43 new arthropod or pathogen agents were released in 19 projects; and that effective biological control was achieved in several projects with the outstanding successes being the control of rubber vine, Cryptostegia grandiflora, and bridal creeper, Asparagus asparagoides.

When is Weed Biological Control Successful?

Information collated on weed impacts before the initiation of a biological control program is necessary to provide baseline data and devise performance criteria with which the program can subsequently be evaluated. For avoidance of confusion on when a biological control could be viewed as successful or not. Hoffmann an implementation of a particular biological control will be termed successful when: complete-when no other control method is required or used, at least in areas where the agent(s) is established; substantial-where other methods are needed but the effort required is reduced (e.g. less herbicide or less frequent application); and negligible-where despite damage inflicted by agents, control of the weed is still dependent on other control measures. Complete control does not imply total eradication of the weed; rather it means that control measures are not required anymore specifically against the target weed, and that crop or pasture yield losses will not be attributed mainly to this weed. Substantial control involves situations where control may be complete in some seasons and over part of the weed's range, as well as cases where the control achieved is widespread and economically significant but the weed is still a major problem. It is therefore concluded that successful implementation of biological approach to weed control is the successful control of the weed, and not necessarily the successful establishment of individual agents released against the weed. Successful biological control depends on three factors: the extent to which each individual agent can limit the targeted plant; the ecology of the agent as it affects its ability to populate and spread easily in the new environment; and the ecology of the weed, which determines if the total damage that can be caused by the agent can significantly reduce its population. Because agents always need some surviving predator plants to complete their life cycle, biological control will not usually totally eradicate their target weeds. In essence a successful biological control program reduces the potency and population of the target weed and usually in conjunction with other control methods as part of an overall integrated weed management scheme which is recommended.

Things to Consider when making the Choice of Agents to be Introduced to Control Weeds

Gassmann selection of potential agents in the last decades has been mainly based on the population biology of the weed, impact studies of agents on the plant and the combined effect of herbivory and plant competition. Palmer et al. agent selection is highly dependent on the type of weed, its reproductive system, on the ecological, abiotic and management context in which that weed occurs, and on the acceptable goals and impact thresholds required of a biological control program. Generally, factors to be considered in selecting agents include the following: the agent must target a particular plant species, must have high level of predation and parasitism on the host plant and its entire population, must be prolific, must be able to thrive in all habitats and climates where the weed exists and should be able to spread easily and widely, must be a strong colonizer, the overall cost of introducing the agent must be cheaper compared to other control methods, the technology that will be involved in introducing and managing the agent must be as simple as possible, must as much as possible maintain natural biodiversity, sufficient number of individuals must be released, plant phenology (effect of periodic plant life cycle events) must be favorable. To be considered a good candidate for biological control, a weed should be non-native, present in numbers and densities greater than in its native range and numerous enough to cause environmental or economic damage, the weed should also be present over a broad geographic range, have few or no redeeming or beneficial qualities, have taxonomic characteristics sufficiently distinct from those of economically important and native plant. Furthermore, the weed should occur in relatively undisturbed areas to allow for the establishment of biological control agents, cultivation, mowing and other disturbances can have a destructive effect on many arthropod biocontrol agents. Inundative biocontrol agents such as bacteria and fungi are less sensitive to these types of disturbances so may be used in cropland.

Steps to Identifying and Introducing Biological Control Agents

The study of insect attributes and fitness traits, the influence of plant resources on insect performance, and the construction of comparative life-tables, are the first steps towards an improvement of the success rate of biological weed control. Generally, steps to identifying and introducing biological control agents include: (i) identify target weeds; (ii) identify control agents and determine the level of specialization; (iii) apply controlled release of the agents; (iv) apply full release and determine optimal release sites; (v) for the case of classical methods, monitor release sites; (vi) apply redistribution for the case of classical methods (vii) and maintain control agent populations.

Ecological Weed Management in Organic Vegetables

Ecological weed management begins with careful planning of the cropping system to minimize weed problems, and seeks to utilize biological and ecological processes in the field and throughout the farm ecosystem to give crops the advantage over weeds. In addition, mechanical and other control measures are usually needed to protect organic crops from the adverse effects of weeds. This is particularly true in vegetables and other annual crops, for which production practices keep natural plant succession at its earliest stages, thereby eliciting the emergence of pioneer plants that can become agricultural weeds.

While tillage and cultivation can degrade soil quality and increase the risk of erosion losses, many other organic weed management tools are more soil-friendly. For example, a diversified rotation of vigorous cash crops and cover crops can enhance soil organic matter, tilth, and fertility, provided that a sufficient quantity and diversity of residues are returned to the soil to feed the soil life. Grazing livestock after a production crop to remove weeds or interdict weed seed set can add fertility in the form of manure, though intensive grazing can also compact the soil. In the interest of food safety, care must be taken to avoid direct contact of fresh manure with vegetables and other food crop. Mowing and flame weeding (if properly done to avoid excessive heating of the soil itself) are much easier on soil structure than cultivation, and can be just as effective in certain stages of weed and crop development. Mowing or rolling a cover crop to form an *in situ* mulch can enhance the soil benefits of the cover crop, compared to tilling it in, and can effectively suppress many annual weeds. Other organic mulches, such as straw and chipped brush, add organic matter, whereas synthetic clear or colored plastic films and weed barrier fabrics do not. All mulches are very effective in preventing soil erosion.

A summary of organic weed management tools.		
	Preventive	Control
Major tools		
The Grower's Mind (planning, observation, and ingenuity)	X	X
Vigorous Cash Crops	X	
Crop Rotation	X	
Cover Crops	X	
Organic Mulches	X	X
Opaque Synthetic Mulches (black plastic, etc.)	X	X
Conservation Biological Control (conserve weed consumers present on farm)	X	X
Livestock	X	X
Tillage and Cultivation Tools and Implements		X
Mowers and other Cutting Tools		X
Rollers and Roll-crimpers (for converting mature cover crops into in-situ mulch)	X	X
Flame Weeders		X
Minor and experimental tools		
OMRI certified organic herbicides		X
Bioherbicides (specific pathogens of weeds)		X
Management of soil microflora	X	X
Specific crop–weed allelopathic interactions	X	X
Classical biological controls for specific weeds (usually against invasive exotic weeds in rangeland and natural ecosystems)		X
Clear plastic mulch (soil solarization)	X	X

Ecological weed management consists of many-component strategies tailored to each region, cropping system, and farm. Matt Liebman and Eric Gallandt describe the process as using "many little hammers", including "indirect controls", such as crop variety, planting date, and nutrient management, rather than relying only on the "direct controls" or "large hammers" of cultivation and herbicides. In their words, "the use of a combination of methods can lead to: (i) acceptable control through the additive, synergistic, or cumulative action of tactics that may not be effective when used alone, (ii) reduced risk of crop failure or serious loss by spreading the burden of protection across several methods, and (iii) minimal exposure to any one tactic, and consequently reduced rates at which pests adapt and become resistant."

Note that these steps do not comprise a precise linear sequence of instructions; rather they offer a conceptual framework within which each farmer can develop a site-specific strategy. This process requires systems thinking, which views the field as a complex system of interacting components—such ascrops, weeds, soil, insects, and microorganisms—that form a web of relationships, not a linear sequence of cause-and-effect. Similarly, the following steps are employed together in a synergistic manner, and thus differ from the sequence of instructions for assembling a car or a farm implement.

For example, Step 6 (cover crops) can be seen as a part of Step 2 (minimize niches for weeds), and Step 1 (know the weeds) provides vital information for other steps, particularly steps 3 (keep the weeds guessing), 4 (design for effective weed control), and 7 (manage the weed seed bank). Biological processes (Step 9) include indigenous biocontrols that help reduce the weed seed bank (Step 7) as well as the competitive and allelopathic effects of cover crops (Step 6). Step 11 (observe weeds and adapt practices) is an ongoing feedback loop that informs and fine-tunes all the other steps. Utilizing this or another suitable framework, the organic grower selects and assembles a set of "many little hammers" that, working together, keep the farm's weeds from becoming major weed problems.

Planning Steps

Know the Weeds

Obtain correct identification of the major weeds present on the farm. Monitor fields regularly throughout the season. Keep records on what weeds emerge at different seasons, and on efficacy of any preventive and control measures taken. Learn each weed's life cycle, growth habit, seasonal pattern of development and flowering, modes of reproduction and dispersal, seed dormancy and germination, and how the weed affects crop production. Find the weed's weak points—possibly the stages in its life cycle that are most vulnerable to control tactics—and stresses to which the weed is sensitive; these can be exploited in designing a management strategy.

"Know the weeds" is listed first because it informs most of the succeeding steps. However, gaining a thorough knowledge of the farm's weed flora is an ongoing process over many seasons (perhaps the lifetime of the farmer!) that drives the year-to-year refinement of the farm's weed management system.

Design the Cropping System to Minimize

Niches for Weed Growth In planning the crop rotation, avoid creating open niches in time or space. Plan tight rotations that follow one crop harvest promptly with the next planting. Open niches in space between crop rows can be reduced by using a narrower row spacing, intercropping, relay cropping, overseeding cover crops into established vegetables, or no-till management of cover crops prior to transplanting vegetables.

Keep the Weeds Guessing with Crop Rotations

Plan and implement diversified crop rotations that vary timing, depth, frequency, and methods of tillage; timing and methods of planting, cultivation, and harvest; as well as crop plant family. Alternate warm- and cool-season vegetables. Rotate vegetable fields into perennial cover for two or three years to interrupt life cycles of annual weeds adapted to frequent tillage. Schedule tillage and cultivation operations when they will do the most damage to the major weed species.

Design the Cropping System and Select Tools for Effective Weed Control

Develop control strategies to address anticipated weed pressures in each of the farm's major crops. Choose the best cultivation implements and other tools for cost-effective preplant, between-row, and within-row weed removal. Plan bed layout, as well as row- and plant spacing, to facilitate precision cultivation. Choose irrigation methods and other cultural practices that are compatible with planned weed control operations.

Preventive Steps during the Season

Grow vigorous, Weed-competitive Crops

A healthy, fast-growing crop that can outcompete weeds is the best way to prevent weed problems. Choose locally-adapted crop varieties that grow tall or form lots of foliage that can shade out weeds. Maintain healthy, living soil. Provide optimum growing conditions—planting date and spacing, moisture, soil tilth and aeration, fertility, and pest and disease management. Deliver water and fertilizer within-row to feed the crop and not the weeds. Note that either insufficient or excessive levels of major nutrients (nitrogen, phosphorus, and potassium) can give certain weeds a competitive advantage over the crops.

Put the Weeds out of Work

Cover crops do the same job as weeds, only better. They rapidly occupy open niches, protect and restore the soil, provide beneficial habitat, add organic matter, and hold and recycle soil nutrients. They suppress weeds through direct competition and sometimes through allelopathy—the release of plant-growth-inhibiting substances into the soil. Whenever a bed or field becomes vacant, plant a cover crop immediately so that it can begin the vital restorative work that nature accomplishes with pioneer plants or weeds. Good cover cropping plays a major role in Step 2 (minimzing open niches), and can put the weeds out of a job.

Manage the Weed Seed Bank minimize Deposits and maximize Withdrawals

Prevent formation and release of viable weed seeds, and proliferation of rhizomes and other propagules of perennial weeds. Avoid importing new weeds with manure, mulch hay, and other materials from off-farm sources. Utilize stale seedbed, cultivated fallow, or targeted tillage practices to draw down seed banks of the major weeds present. Encourage weed seed mortality and weed seed consumption by ground beetles and other organisms.

Control Steps during the Season

Knock Out Weeds at Critical Times

Plant vegetables into a clean seedbed, hit early-season weeds while the are small, and keep crops clean through their critical weed free period (through the first third or half of the life cycle of most vegetables). Prevent seed set by "escapes" and late season weeds. When practical, interrupt vegetative propagation by invasive perennial weeds through timely removal of top growth.

Utilize Biological Control Processes to Further Reduce Weed Pressure

Rotate livestock, poultry, or weeder geese through fields to graze weeds and interrupt seed set. To ensure food safety and comply with USDA Organic Standards, time such grazing so that fresh droppings are not deposited any less than 120 days prior to harvest of the next crop. Encourage weed seed predation and decay by maintaining high soil biological activity and providing habitat (mulch, cover crops, hedgerows) for belowground and aboveground weed seed consumers (conservation biological control). Enhance overall soil biological activity to tip the competitive balance in favor of crops, and possibly to shorten the "half life" of the weed seed bank.

Classical biological controls (introduced natural enemies) are commercially available for a few invasive exotic weeds.

Bring existing Weeds under Control before Planting Weed-sensitive Crops

Weed control in perennial horticultural crops like asparagus, small fruit, and some cut flowers can be quite difficult, especially when perennial weeds dominate the weed flora. Bring existing weed pressures under good control through repeated tillage and intensive cover cropping before planting any perennial vegetable, fruit, or ornamental crops. Choose fields with the best weed control or lowest weed pressure for weed-sensitive annual vegetables with a long critical weed free period, such as carrot, onion,and parsnip. Be sure weeds, especially perennial weeds, are under good control before attempting no-till management of cover crops prior to cash crop planting.

Enhancing the Organic Weed Management System: Observe, Adapt and Experiment

Keep observing the weeds and adapt practices accordingly, note and record any changes in weed species composition, emergence and growth pattern, or weed pressure, and modify practices as needed. For example, an increase in certain annual "weeds of cultivation" may indicate a need to reduce tillage or diversify the crop rotation. An increase in invasive perennials may require tilling deeper or more aggressively for a time. Watch out for the arrival of new weed species that could pose problems.

Expect weed populations and flora to shift over time. Every farm decision and field operation can elicit changes in the weed community, as can weather variations, to say nothing of long term climate changes. "Reading" the weeds each year becomes an information feedback loop, guiding weed management practices for the following season.

Experiment: Try out new tactics and strategies to deal with major weed challenges. Farmers continually develop innovative strategies based on new tools that they fashion themselves or that researchers develop, new uses for old tools, and new combinations of preventive and control tactics. In the words of University of Vermont Extension Specialist Vern Grubinger.

Experiment to fine-tune your weed management tactics:

- Start on a small scale with tools and techniques that are new to your farm.
- Identify your most problematic weeds and compare different combinations or rotations, cover crops, and cultivation tools to see how effective they are in providing control.
- Keep an eye out for new tools, or new ways to use old tools.
- Leave a control row or section untreated, so you can see the effectiveness of your tactics.

Part of experimenting is to watch for new developments. Researchers and farmers continue to explore and expand the horizons of possibility in ecological and organic weed management. Check farming magazines and publications for practical applications of their work, from new cultivation tools to new strategies for particularly stubborn weed problems. Some cutting edge areas of research may take longer to yield practical results, yet bear watching and possible integration into a farm's weed management strategy. These range from natural herbicides, bioherbicides, and classical biological controls, to specific weed–crop allelopathic interactions and manipulation of weed–crop–soil–microbe relationships to give the crop a competitive edge over certain weeds.

While none of these endeavors is likely to yield a "big hammer" to replace herbicides or steel, they can contribute additional "little hammers" to enhance efficacy and reduce the amount of tillage and cultivation needed.

An effective organic weed management system cannot be spelled out precisely because ecological weed management is inherently site specific and responsive to changes in the farm ecosystem. There is effectively no "organic weed control cookbook" to replace the precise herbicide protocols that have been developed for conventional production of row crops and some vegetables. No scientist can come up with a better weed management strategy for a particular farm than the strategy a skillful organic farmer can develop by applying ecological weed management principles to the particular suite of crops, weeds, soil conditions, and available resources on her or his farm.

References

- Lemaux, Peggy G. (2008). "Genetically Engineered Plants and Foods: A Scientist's Analysis of the Issues (Part I)". Annual Review of Plant Biology. 59: 771–812. Doi:10.1146/annurev.arplant.58.032806.103840. PMID 18284373
- Ecological-pest-management, content: climatetechwiki.org, Retrieved 26 June, 2019
- Classical Biological Control: Importation of New Natural Enemies". University of Wisconsin. Archived from the original on 13 June 2016. Retrieved 7 June 2016
- Bioherbicides, herbicides-physiology-of-action-and-safety, books: intechopen.com, Retrieved 28 August, 2019
- Peter, K. V. (2009). Basics Of Horticulture. New India Publishing. P. 288. ISBN 978-81-89422-55-4. Archived from the original on 2017-04-07
- Biological-pest-control-uses-advantages-and-disadvantages, biology: online-sciences.com, Retrieved 27 July, 2019
- "With biodirect, Monsanto Hopes RNA Sprays Can Someday Deliver Drought Tolerance and Other Traits to Plants on Demand | MIT Technology Review". Retrieved 2015-08-31
- Biological-control, weed, wssa: net, Retrieved 29 January, 2019

5

Allied Fields of Agroecology

There are various fields which are closely associated with agroecology such as agroforestry, sustainable agriculture, carbon farming, sustainable gardening, permaculture, polyculture, silviculture and organic farming. The topics elaborated in this chapter will help in gaining a better perspective about these allied fields of agroecology.

SUSTAINABLE AGRICULTURE

Sustainable agriculture is a type of agriculture that focuses on producing long-term crops and livestock while having minimal effects on the environment. This type of agriculture tries to find a good balance between the need for food production and the preservation of the ecological system within the environment. In addition to producing food, there are several overall goals associated with sustainable agriculture, including conserving water, reducing the use of fertilizers and pesticides, and promoting biodiversity in crops grown and the ecosystem. Sustainable agriculture also focuses on maintaining economic stability of farms and helping farmers improve their techniques and quality of life.

Sustainable agriculture, in a simple way, is "capable of maintaining productivity and usefulness to society indefinitely". In other words, it is resource-conserving, socially supportive, commercially competitive, and environmentally sound. Under the law, the term sustainable agriculture means an integrated system of plant and animal production practices having a site-specific application and such a system over the long term satisfies human food and fibber needs, enhances environmental quality and the natural resource base upon which the agricultural economy depends, makes the most efficient use of non renewable resources and on-farm resources and integrate, where appropriate, natural biological cycles and controls, sustain the economic viability of farm operations and enhance the quality of life for farmers and society as a whole.

Therefore, sustainable agriculture does not mean a return to either the low yields or poor farmers that characterized the 19th century. Rather, sustainability builds on current agricultural achievements, adopting a sophisticated approach that can maintain high yields and farm profits without undermining the resources on which agriculture depends.

Sustainable agriculture refers to the ability of a farmland to produce food perpetually. It involves two key issues long-term effects of various practices on soil properties and processes essential for crop productivity, and the long-term availability of inputs. Sustainable agriculture integrates three main goals environmental health, economic profitability and social economic equity. A variety of philosophies, policies and practices have contributed to these goals. People in many different capacities from farmers to consumers have shared this vision and contributed to it.

Sustainability rests on the principle that we must meet the needs of the present without compromising the ability of future generations to meet their own needs. Therefore, stewardship of both natural and human resources is of prime importance. Stewardship of human resources includes consideration of social responsibilities such as working and living conditions of labourers, the needs of rural communities, and consumer health and safety both in the present and the future.

Stewardship of land and natural resources involves maintaining or enhancing this vital resource base for the long term. A systems approach is required to understand and achieve sustainability. It gives us the tools to explore the interconnections between farming and other aspects of our environment. It requires not only the input of researchers from various disciplines, but also farmers, farm workers, consumers, policymakers and others.

Therefore, sustainable agriculture addresses many environmental and social concerns and offers innovative and economically viable opportunities for growers, labourers, consumers, policymakers and many others in the entire food system. The key to reducing world hunger, poverty and the harmful environmental effects of both industrialized and traditional agriculture is to develop a variety of systems of sustainable agriculture.

Each region has a unique set of agro-ecosystems that result from local variations in climate, soil, economic relation, social structure and history. Keeping this in view, sustainable agricultural practices have to be designed and implemented in different eco-regions of any country.

In sustainable agriculture, the soil is viewed as a fragile and living medium that must be protected and nurtured to ensure its long-term productivity and stability. Further, it recognizes that a "healthy" soil is a key component as it is the base to produce healthy crop plants that have optimum vigour and are less susceptible to pests. Methods such as cover crops, compost and manures, reducing tillage, avoiding traffic on wet soils, and maintaining soil cover with plants and mulches, if followed, protect and enhance the productivity of the soil.

The land requires replenishment from time to time without which the land would suffer from nutrient depletion and becomes unusable for further farming. Sustainable agriculture depends on replenishing the soil while minimizing the use of non-renewable resources, such as natural gas in converting atmospheric nitrogen into synthetic fertilizer or mineral ores like phosphate. Rainfall is an important component; in some areas, sufficient rainfall is available for crop growth while many areas require irrigation. In such areas, irrigation systems should be sustainable, managed properly to avoid salt accumulation and use water from the source rationally.

Sustainable agriculture allows growing crops and raising livestock based on organic fertilizers, soil and water conservation, and biological control of pests and minimal use of non-renewable fossil fuel energy. It speaks about the ability of a farm to continue production indefinitely with a minimum of outside inputs. Crops depend on nutrients from soil, air, water, and sunlight to product the foodstuff that human beings need to live.

When farmers harvest crops, they take what crops have produced from the resources available to them. These resources must be replenished to allow the production cycle to continue. Otherwise, the resources would be exhausted and the land would be unusable for further farming. Although resources like the sun, air, and rain are generally available in most geographic locations, nutrients in the soil are easily exhausted.

Adding off-farm inputs, such as fertilizer for plants, or petroleum products to run machinery, reduce sustainability due to a reliance on non-renewable resources. The fewer outside inputs the farm needs to maintain production levels, the greater is its sustainability. Nutrients in the soil may be replenished through recycling of crop residues and livestock manure with their nutrients into the soil. Labour by animals or farmers is another form of energy recycling, if they are fed with the food grown and harvested from the farm.

From an environmental perspective, given the finite supply of natural resources, agriculture that is inefficient and low on the sustainability scale will eventually exhaust the available resources, or the ability to afford them, and cease to be viable as a farming method. It will also generate negative externality, an economic term for by-products of production, such as pollution, financial and production costs. Agriculture that relies mainly on inputs, which are extracted from the Earth's crust or produced by the society, contributes to the depletion and degradation of the environment.

In an economic context, the farm must generate revenue to acquire things that cannot be produced directly. The way that crops are sold must then be accounted for in the sustainability equation. Fresh food sold from a farm stand requires little additional energy, beyond cultivation and harvest, though the cost of consumers' transport to the site must be included.

Food that is packaged and sold at a remote location, such as farmers' market, incurs a greater energy cost for materials, labour, transport, and so forth. A more complex economic system in which the farm producer is just the first link in a long chain of processors and handlers leads to greater costs and greater reliance on the use of externals. Such a system is vulnerable to fluctuation in prices of imported external materials.

Sustainable development requires a variety of approaches. Specific strategies must take into account the aspects like topography, soil characteristics, climate, pest, local availability of inputs and the individual grower goals. Despite the site specific and individual nature of sustainable agriculture, several general principles can be applied to help grower's select appropriate management practices such as selection of species and varieties that are well-suited to the site and to the condition of the farm and diversification of crops and cultural practices to enhance the biological and economic stability of the farm, and manage soil to enhance and protect the soil quality.

Sustainable environmental management in agriculture ensures a sound balance between optimum agricultural productivity and renewal of natural resources. This is possible only if the pragmatic principles of ecology are properly adhered to in planning, management and developmental activities in agriculture, giving due care to agro-chemical and socio-economic needs.

Sustainable farming practices include:

1. Crop rotations that mitigate weeds, disease, insect and other pest problems; provide alternative sources of soil nitrogen; reduce soil erosion; and reduce risk of water contamination by agricultural chemicals.

2. Pest control strategies that are not harmful to natural systems, farmers, their neighbours, or consumers. This includes integrated pest management techniques that reduce the need for pesticides by practices such as scouting, use of resistant cultivars, timing of planting, and biological pest controls.
3. Increased mechanical/biological weed control; more soil and water conservation practices; and strategic use of animal and green manures.
4. Use of natural or synthetic inputs in a way that poses no significant hazard to man, animals, or the environment.

Sustainable agriculture is a model of social and economic organization based on equitable and participatory vision of development which recognizes the environment and natural resources as the foundation of economic activity. Agriculture is sustainable when it is ecologically sound, economically viable, socially just, culturally appropriate and based on a holistic scientific approach.

It preserves biodiversity, maintains soil fertility and water purity, conserves and improves the chemical, physical and biological qualities of the soil, recycles natural resources and conserves energy. It produces diverse forms of high quality foods, fibbers and medicines.

It uses locally available renewable resources, appropriate and affordable technologies and minimizes the use of external and purchased inputs, thereby increasing local independence and self-sufficiency and ensuring a source of stable income for peasants' family and small farmers and rural communities.

This allows more people to stay on the land, strengthens rural communities and integrates humans with their environment. Sustainable agriculture respects the ecological principles of diversity and interdependence and uses the insights of modern science to improve rather than displace the traditional wisdom accumulated over centuries by innumerable farmers around the world.

It does not refer to a prescribed set of practices. Instead, it challenges producers to think about the long-term implications of practices and the broad interactions and dynamics of agricultural systems. It also invites consumers to get more involved in agriculture by learning more about and becoming active participants in their food systems.

Finally, it is a whole-systems approach to food, feed and other fibber production that balances environmental soundness, social equity, and economic viability among all sectors of the public, including international and intergenerational peoples. Inherent in this definition is the idea that sustainability must be extended not only globally, but indefinitely in time, and to all living organisms including humans.

Therefore, sustainable agro-ecosystems maintain their natural resource base, rely on minimum artificial inputs from outside the farm system, manage pests and diseases through internal regulating mechanisms and recover from the disturbances caused by cultivation and harvest. In view of the many negative effects of modern agriculture, which have far-reaching consequences on the environment, sustainable agriculture has become an integral component of many government, commercial, and non-profit agriculture research efforts.

Sustainable agriculture concept is a big threat to people involved in agribusiness and to successful farmers with large investments in industrialized agriculture, and to specialized farmers unwilling to learn the demanding art of fanning sustainably. Further, it may receive resistance from many consumers unwilling or unable to pay higher prices for food, because full cost accounting would include agriculture's harmful environmental and health costs in the market prices of food.

Despite these difficulties, environmentalists believe that a shift from modern agriculture to sustainable agriculture could be brought about over the next 30-50 years by instituting several policies with some subsidies and tax breaks.

AGROFORESTRY

Agroforestry is the cultivation and use of trees and shrubs with crops and livestock in agricultural systems. Agroforestry seeks positive interactions between its components, aiming to achieve a more ecologically diverse and socially productive output from the land than is possible through conventional agriculture. Agroforestry is a practical and low-cost means of implementing many forms of integrated land management (which seeks to reduce human impacts on land), and it contributes to a green economy by promoting long-term, sustainable, and renewable forest management, especially for small-scale producers. Although the modern concept of agroforestry emerged in the early 20th century, the use of woody perennials in agricultural systems is ancient, with written descriptions of the practice dating back to Roman times. Indeed, integrating trees with crops and animals is a long-standing tradition throughout the world. In 2004, the World Bank estimated that agroforestry practices were being used by 1.2 billion people.

Irrigated rice terraces.

Benefits of Agroforestry

Agroforestry can occur at a variety of spatial scales (e.g., field or woodlot, farm, watershed) in different ecosystems and cultures. When properly applied, agroforestry can improve livelihoods through enhanced health and nutrition, increased economic growth, and strengthened environmental resilience and ecosystem sustainability. In turn, such improvements can contribute to increased social sustainability in which human needs are satisfied in a way that fosters environmental health. Farm diversification is a growing strategy for economic competitiveness, especially throughout the industrialized temperate zone, and agroforestry offers great promise for the sustainable production of specialty nut and fruit crops, high-value medicinals, dairy and beef cattle, sheep, goats, and biomass for biofuel. Agroforestry systems also yield proven strategies for long-term carbon sequestration, soil enrichment, biodiversity conservation, and air- and water-quality improvements, benefiting both the landowners and society.

The benefits of agroforestry derive from the interactions between trees and shrubs and crops and livestock. Agroforestry seeks to optimize positive interactions, such as mutualism and commensalism, and to minimize predation on crops and livestock and competition within and between species. Positive interactions may reduce stress to plants and animals, enhance yields, retain soil, and capture water. For example, the moist shaded microclimate under certain crop trees is beneficial for shade-tolerant crops such as turmeric or pineapple. Negative interactions, by contrast, can result in resource competition, more pests, excessive shading, and allelopathy (the release of biochemicals by one plant to suppress the growth of another). Black walnut and various eucalyptus trees, for example, are known to allelopathically inhibit the growth of certain annual crops planted near them.

Agroforestry Practices

Agroforestry systems are intensively managed to maintain their productive and protective functions through cultivation, fertilization, irrigation, pruning, and thinning. Ideally, components are structurally and functionally combined and actively managed to optimize the positive biophysical interactions between them. In some systems, for example, the trees are regularly coppiced (severely cut back), and the cuttings are applied as mulch to the soil. Such management not only encourages new tree growth but also augments the light levels reaching shaded crops, reduces weeds, and helps to maintain soil moisture.

Although American and Canadian temperate-zone agroforestry nomenclature differs from that used in the tropics and Europe, five temperate-zone agroforestry practices are generally recognized worldwide.

- Riparian and upland buffers: Strips of permanent vegetation, consisting of trees, shrubs, grasses, and forbs planted and managed together.
- Windbreaks: Trees or shrubs planted and managed as barriers to reduce wind speed as part of a crop or livestock operation.
- Alley cropping (known as silvoarable agroforestry in Europe): Trees planted in multiple rows combined with crops cultivated in the alleyways between the tree rows.
- Silvopasture (also known as agrosilvopastoral agroforestry or *dehesa*): trees combined with forage (pasture) and livestock production.
- Forest farming: Cultivation of high-value specialty crops under the protection of a forest overstory that provides an appropriate microclimate.

In addition to many of the temperate-zone practices, tropical agroforestry systems often include a variety of other agroforestry methods. Home gardens and taungya, in which food crops are grown between tree seedlings as they mature (often used in teak or mahogany production), are useful to supplement the nutritional needs of landowners growing tree crops. Terrace cultivation, living fences, multipurpose trees (e.g., the Albida acacia [Faidherbia albida]), fodder trees, and multi-strata systems (featuring trees of different heights, as in shade-grown coffee) are also used in tropical systems to combine trees with other crops and livestock.

ORGANIC FARMING

Organic farming is an alternative agricultural system which originated early in the 20th century in reaction to rapidly changing farming practices. Certified organic agriculture accounts for 70 million hectares globally, with over half of that total in Australia. Organic farming continues to be developed by various organizations today. It is defined by the use of fertilizers of organic origin such as compost manure, green manure, and bone meal and places emphasis on techniques such as crop rotation and companion planting. Biological pest control, mixed cropping and the fostering of insect predators are encouraged. Organic standards are designed to allow the use of naturally occurring substances while prohibiting or strictly limiting synthetic substances. For instance, naturally occurring pesticides such as pyrethrin and rotenone are permitted, while synthetic fertilizers and pesticides are generally prohibited. Synthetic substances that are allowed include, for example, copper sulfate, elemental sulfur and Ivermectin. Genetically modified organisms, nanomaterials, human sewage sludge, plant growth regulators, hormones, and antibiotic use in livestock husbandry are prohibited. Reasons for advocation of organic farming include advantages in sustainability, openness, self-sufficiency, autonomy/independence, health, food security, and food safety.

World map of organic agriculture (hectares).

Organic agricultural methods are internationally regulated and legally enforced by many nations, based in large part on the standards set by the International Federation of Organic Agriculture Movements (IFOAM), an international umbrella organization for organic farming organizations established in 1972. Organic agriculture can be defined as:

> An integrated farming system that strives for sustainability, the enhancement of soil fertility and biological diversity whilst, with rare exceptions, prohibiting synthetic pesticides, antibiotics, synthetic fertilizers, genetically modified organisms, and growth hormones.

Since 1990, the market for organic food and other products has grown rapidly, reaching $63 billion worldwide in 2012. This demand has driven a similar increase in organically managed farmland that grew from 2001 to 2011 at a compounding rate of 8.9% per annum.

Vegetables from ecological farming.

As of 2019, approximately 70,000,000 hectares (170,000,000 acres) worldwide were farmed organically, representing approximately 1.4 percent of total world farmland.

Methods

Organic cultivation of mixed vegetables in Capay, California.

> Organic agriculture is a production system that sustains the health of soils, ecosystems and people. It relies on ecological processes, biodiversity and cycles adapted to local conditions, rather than the use of inputs with adverse effects. Organic agriculture combines tradition, innovation and science to benefit the shared environment and promote fair relationships and a good quality of life for all involved.

Organic farming methods combine scientific knowledge of ecology and some modern technology with traditional farming practices based on naturally occurring biological processes. Organic farming methods are studied in the field of agroecology. While conventional agriculture uses synthetic pesticides and water-soluble synthetically purified fertilizers, organic farmers are restricted by regulations to using natural pesticides and fertilizers. An example of a natural pesticide is pyrethrin, which is found naturally in the Chrysanthemum flower. The principal methods of organic farming include crop rotation, green manures and compost, biological pest control, and mechanical cultivation. These measures use the natural environment to enhance agricultural productivity: legumes are planted to fix nitrogen into the soil, natural insect predators are encouraged, crops are rotated to confuse pests and renew soil, and natural materials such as potassium bicarbonate and mulches are used to control disease and weeds. Genetically modified seeds and animals are excluded.

While organic is fundamentally different from conventional because of the use of carbon based fertilizers compared with highly soluble synthetic based fertilizers and biological pest control instead of synthetic pesticides, organic farming and large-scale conventional farming are not entirely mutually exclusive. Many of the methods developed for organic agriculture have been borrowed by more conventional agriculture. For example, Integrated Pest Management is a multifaceted strategy that uses various organic methods of pest control whenever possible, but in conventional farming could include synthetic pesticides only as a last resort.

Crop Diversity

Organic farming encourages crop diversity. The science of agroecology has revealed the benefits of polyculture (multiple crops in the same space), which is often employed in organic farming. Planting a variety of vegetable crops supports a wider range of beneficial insects, soil microorganisms, and other factors that add up to overall farm health. Crop diversity helps environments thrive and protects species from going extinct.

Soil Management

Organic farming relies heavily on the natural breakdown of organic matter, using techniques like green manure and composting, to replace nutrients taken from the soil by previous crops. This biological process, driven by microorganisms such as mycorrhiza, allows the natural production of nutrients in the soil throughout the growing season, and has been referred to as feeding the soil to feed the plant. Organic farming uses a variety of methods to improve soil fertility, including crop rotation, cover cropping, reduced tillage, and application of compost. By reducing tillage, soil is not inverted and exposed to air; less carbon is lost to the atmosphere resulting in more soil organic carbon. This has an added benefit of carbon sequestration, which can reduce green house gases and help reverse climate change.

Plants need a large number of nutrients in various quantities to flourish. Supplying enough nitrogen and particularly synchronization so that plants get enough nitrogen at the when plants need it most, is a challenge for organic farmers. Crop rotation and green manure ("cover crops") help to provide nitrogen through legumes (more precisely, the family Fabaceae), which fix nitrogen from the atmosphere through symbiosis with rhizobial bacteria. Intercropping, which is sometimes used for insect and disease control, can also increase soil nutrients, but the competition between the legume and the crop can be problematic and wider spacing between crop rows is required. Crop residues can be ploughed back into the soil, and different plants leave different amounts of nitrogen, potentially aiding synchronization. Organic farmers also use animal manure, certain processed fertilizers such as seed meal and various mineral powders such as rock phosphate and green sand, a naturally occurring form of potash that provides potassium. Together these methods help to control erosion. In some cases pH may need to be amended. Natural pH amendments include lime and sulfur, but in the U.S. some compounds such as iron sulfate, aluminum sulfate, magnesium sulfate, and soluble boron products are allowed in organic farming.

Mixed farms with both livestock and crops can operate as ley farms, whereby the land gathers fertility through growing nitrogen-fixing forage grasses such as white clover or alfalfa and grows cash crops or cereals when fertility is established. Farms without livestock ("stockless") may find it more difficult to maintain soil fertility, and may rely more on external inputs such as imported manure as well as grain legumes and green manures, although grain legumes may fix limited nitrogen because they are harvested. Horticultural farms that grow fruits and vegetables in protected conditions often rely even more on external inputs.

Biological research into soil and soil organisms has proven beneficial to organic farming. Varieties of bacteria and fungi break down chemicals, plant matter and animal waste into productive soil nutrients. In turn, they produce benefits of healthier yields and more productive soil for future crops. Fields with less or no manure display significantly lower yields, due to decreased soil microbe community. Increased manure improves biological activity, providing a healthier, more arable soil system and higher yields.

Weed Management

Organic weed management promotes weed suppression, rather than weed elimination, by enhancing crop competition and phytotoxic effects on weeds. Organic farmers integrate cultural, biological, mechanical, physical and chemical tactics to manage weeds without synthetic herbicides.

Organic standards require rotation of annual crops, meaning that a single crop cannot be grown in the same location without a different, intervening crop. Organic crop rotations frequently include weed-suppressive cover crops and crops with dissimilar life cycles to discourage weeds associated with a particular crop. Research is ongoing to develop organic methods to promote the growth of natural microorganisms that suppress the growth or germination of common weeds.

Other cultural practices used to enhance crop competitiveness and reduce weed pressure include selection of competitive crop varieties, high-density planting, tight row spacing, and late planting into warm soil to encourage rapid crop germination.

Mechanical and physical weed control practices used on organic farms can be broadly grouped as:

- Tillage - Turning the soil between crops to incorporate crop residues and soil amendments; remove existing weed growth and prepare a seedbed for planting; turning soil after seeding to kill weeds, including cultivation of row crops;
- Mowing and cutting - Removing top growth of weeds;
- Flame weeding and thermal weeding - Using heat to kill weeds; and
- Mulching - Blocking weed emergence with organic materials, plastic films, or landscape fabric. Some critics, citing work published in 1997 by David Pimentel of Cornell University, which described an epidemic of soil erosion worldwide, have raised concerned that tillage contribute to the erosion epidemic. The FAO and other organizations have advocated a 'no-till' approach to both conventional and organic farming, and point out in particular that crop rotation techniques used in organic farming are excellent no-till approaches. A study published in 2005 by Pimentel and colleagues confirmed that 'Crop rotations and cover cropping (green manure) typical of organic agriculture reduce soil erosion, pest problems, and pesticide use.'

Some naturally sourced chemicals are allowed for herbicidal use. These include certain formulations of acetic acid (concentrated vinegar), corn gluten meal, and essential oils. A few selective bioherbicides based on fungal pathogens have also been developed. At this time, however, organic herbicides and bioherbicides play a minor role in the organic weed control toolbox.

Weeds can be controlled by grazing. For example, geese have been used successfully to weed a range of organic crops including cotton, strawberries, tobacco, and corn, reviving the practice of keeping cotton patch geese, common in the southern U.S. before the 1950s. Similarly, some rice farmers introduce ducks and fish to wet paddy fields to eat both weeds and insects.

Controlling other Organisms

Chloroxylon is used for Pest Management in Organic Rice Cultivation in Chhattisgarh, India.

Organisms aside from weeds that cause problems on organic farms include arthropods (e.g., insects, mites), nematodes, fungi and bacteria. Organic practices include, but are not limited to:

- Encouraging predatory beneficial insects to control pests by serving them nursery plants and an alternative habitat, usually in a form of a shelterbelt, hedgerow, or beetle bank;
- Encouraging beneficial microorganisms;
- Rotating crops to different locations from year to year to interrupt pest reproduction cycles;
- Planting companion crops and pest-repelling plants that discourage or divert pests;
- Using row covers to protect crops during pest migration periods;
- Using biologic pesticides and herbicides;
- Using stale seed beds to germinate and destroy weeds before planting;
- Using sanitation to remove pest habitat;
- Using insect traps to monitor and control insect populations; and
- Using physical barriers, such as row covers.

Examples of predatory beneficial insects include minute pirate bugs, big-eyed bugs, and to a lesser extent ladybugs (which tend to fly away), all of which eat a wide range of pests. Lacewings are also effective, but tend to fly away. Praying mantis tend to move more slowly and eat less heavily. Parasitoid wasps tend to be effective for their selected prey, but like all small insects can be less effective outdoors because the wind controls their movement. Predatory mites are effective for controlling other mites.

Naturally derived insecticides allowed for use on organic farms use include Bacillus thuringiensis (a bacterial toxin), pyrethrum (a chrysanthemum extract), spinosad (a bacterial metabolite), neem (a tree extract) and rotenone (a legume root extract). Fewer than 10% of organic farmers use these pesticides regularly; one survey found that only 5.3% of vegetable growers in California use rotenone while 1.7% use pyrethrum. These pesticides are not always more safe or environmentally friendly than synthetic pesticides and can cause harm. The main criterion for organic pesticides is that they are naturally derived, and some naturally derived substances have been controversial. Controversial natural pesticides include rotenone, copper, nicotine sulfate, and pyrethrums Rotenone and pyrethrum are particularly controversial because they work by attacking the nervous system, like most conventional insecticides. Rotenone is extremely toxic to fish and can induce symptoms resembling Parkinson's disease in mammals. Although pyrethrum (natural pyrethrins) is more effective against insects when used with piperonyl butoxide (which retards degradation of the pyrethrins), organic standards generally do not permit use of the latter substance.

Naturally derived fungicides allowed for use on organic farms include the bacteria Bacillus subtilis and Bacillus pumilus; and the fungus Trichoderma harzianum. These are mainly effective for diseases affecting roots. Compost tea contains a mix of beneficial microbes, which may attack or out-compete certain plant pathogens, but variability among formulations and preparation methods may contribute to inconsistent results or even dangerous growth of toxic microbes in compost teas. Some naturally derived pesticides are not allowed for use on organic farms. These include nicotine sulfate, arsenic, and strychnine.

Synthetic pesticides allowed for use on organic farms include insecticidal soaps and horticultural oils for insect management; and Bordeaux mixture, copper hydroxide and sodium bicarbonate for managing fungi. Copper sulfate and Bordeaux mixture (copper sulfate plus lime), approved for organic use in various jurisdictions, can be more environmentally problematic than some synthetic fungicides dissallowed in organic farming Similar concerns apply to copper hydroxide. Repeated application of copper sulfate or copper hydroxide as a fungicide may eventually result in copper accumulation to toxic levels in soil, and admonitions to avoid excessive accumulations of copper in soil appear in various organic standards and elsewhere. Environmental concerns for several kinds of biota arise at average rates of use of such substances for some crops. In the European Union, where replacement of copper-based fungicides in organic agriculture is a policy priority, research is seeking alternatives for organic production.

Livestock

Raising livestock and poultry, for meat, dairy and eggs, is another traditional farming activity that complements growing. Organic farms attempt to provide animals with natural living conditions and feed. Organic certification verifies that livestock are raised according to the USDA organic regulations throughout their lives. These regulations include the requirement that all animal feed must be certified organic.

For livestock, like these healthy cows, vaccines play an important part in animal health since antibiotic therapy is prohibited in organic farming.

Organic livestock may be, and must be, treated with medicine when they are sick, but drugs cannot be used to promote growth, their feed must be organic, and they must be pastured.

Also, horses and cattle were once a basic farm feature that provided labor, for hauling and plowing, fertility, through recycling of manure, and fuel, in the form of food for farmers and other animals. While today, small growing operations often do not include livestock, domesticated animals are a desirable part of the organic farming equation, especially for true sustainability, the ability of a farm to function as a self-renewing unit.

Genetic Modification

A key characteristic of organic farming is the rejection of genetically engineered plants and animals. On 19 October 1998, participants at IFOAM's 12th Scientific Conference issued the Mar del Plata Declaration, where more than 600 delegates from over 60 countries voted unanimously to exclude the use of genetically modified organisms in food production and agriculture.

Although opposition to the use of any transgenic technologies in organic farming is strong, agricultural researchers Luis Herrera-Estrella and Ariel Alvarez-Morales continue to advocate integration of transgenic technologies into organic farming as the optimal means to sustainable agriculture, particularly in the developing world. Organic farmer Raoul Adamchak and geneticist Pamela Ronald write that many agricultural applications of biotechnology are consistent with organic principles and have significantly advanced sustainable agriculture.

Although GMOs are excluded from organic farming, there is concern that the pollen from genetically modified crops is increasingly penetrating organic and heirloom seed stocks, making it difficult, if not impossible, to keep these genomes from entering the organic food supply. Differing regulations among countries limits the availability of GMOs to certain countries.

Tools

Organic farmers use a number of traditional farm tools to do farming. Due to the goals of sustainability in organic farming, organic farmers try to minimize their reliance on fossil fuels. In the developing world on small organic farms tools are normally constrained to hand tools and diesel powered water pumps.

Disadvantages

It is difficult to compare the Global warming potential, acidification and eutrophication emissions but "Organic production often results in increased burdens, from factors such as N leaching and N_2O emissions", even though primary energy use was less for most organic products. N_2O is always the largest global warming potential contributor except in tomatoes. However, "organic tomatoes always incur more burdens (except pesticide use)". Some emissions were lower "per area", but organic farming always required 65 to 200% more field area than non-organic farming. The numbers were highest for bread wheat (200+ % more) and potatoes (160% more).

Environmental Impact and Emissions

Researchers observed that organic products are sometimes worse for the environment. Organic milk, cereals, and pork generated higher greenhouse gas emissions per product than conventional ones but organic beef and olives had lower emissions in most studies. Usually organic products required less energy, but more land. Per unit of product, organic produce generates higher nitrogen leaching, nitrous oxide emissions, ammonia emissions, eutrophication, and acidification potential than conventionally grown produce. Other differences were not significant. The researchers concluded that public debate should consider various manners of employing conventional or organic farming, and not merely debate conventional farming as opposed to organic farming. They also sought to find specific solutions to specific circumstances.

Proponents of organic farming have claimed that organic agriculture emphasizes closed nutrient cycles, biodiversity, and effective soil management providing the capacity to mitigate and even reverse the effects of climate change and that organic agriculture can decrease fossil fuel emissions. "The carbon sequestration efficiency of organic systems in temperate climates is almost double (575–700 kg carbon per ha per year – 510–625 lb/ac/an) that of conventional treatment of soils, mainly owing to the use of grass clovers for feed and of cover crops in organic rotations."

Critics of organic farming methods believe that the increased land needed to farm organic food could potentially destroy the rainforests and wipe out many ecosystems.

Nutrient Leaching

Nitrogen leaching, nitrous oxide emissions, ammonia emissions, eutrophication potential and acidification potential were higher for organic products, although in one study "nitrate leaching was 4.4–5.6 times higher in conventional plots than organic plots". Excess nutrients in lakes, rivers, and groundwater can cause algal blooms, eutrophication, and subsequent dead zones. In addition, nitrates are harmful to aquatic organisms by themselves.

Land Use

Organic farming requires 84% more land for an equivalent amount of harvest, mainly due to lack of nutrients but sometimes due to weeds, diseases or pests, lower yielding animals and land required for fertility building crops. While organic farming does not necessarily save land for wildlife habitats and forestry in all cases, the most modern breakthroughs in organic are addressing these issues with success.

Professor Wolfgang Branscheid says that organic animal production is not good for the environment, because organic chicken requires twice as much land as "conventional" chicken and organic pork a quarter more. According to a calculation by Hudson Institute, organic beef requires three times as much land. On the other hand, certain organic methods of animal husbandry have been shown to restore desertified, marginal, and otherwise unavailable land to agricultural productivity and wildlife. Or by getting both forage and cash crop production from the same fields simultaneously, reduce net land use.

In England organic farming yields 55% of normal yields. In other regions of the world, organic methods have started producing record yields.

Pesticides

A sign outside of an organic apple orchard reminding orchardists not to spray pesticides on these trees.

In organic farming synthetic pesticides are generally prohibited. A chemical is said to be synthetic if it does not already exist in the natural world. But the organic label goes further and usually prohibit compounds that exist in nature if they are produced by chemical synthesis. So the prohibition is also about the method of production and not only the nature of the compound.

A non-exhaustive list of organic approved pesticides with their median lethal doses:

- Boric acid is used as an insecticide (LD_{50}: 2660 mg/kg).
- Bromomethane is a gas that is still used in the nurseries of Strawberry organic farming.
- Copper(II) sulfate is used as a fungicide and is also used in conventional agriculture (LD_{50} 300 mg/kg). Conventional agriculture has the option to use the less toxic Mancozeb (LD_{50} 4,500 to 11,200 mg/kg).
- Lime sulfur (aka calcium polysulfide) and sulfur are considered to be allowed, synthetic materials (LD_{50}: 820 mg/kg).
- Neem oil is used as an insect repellant.since it contains azadirachtin its use is restricted in the UK and Europe.

- Pyrethrin comes from chemicals extracted from flowers of the genus Pyrethrum (LD_{50} of 370 mg/kg). Its potent toxicity is used to control insects.
- Rotenone is a powerful insecticide that was used to control insects (LD_{50}: 132 mg/kg). Despite the high toxicity of Rotenone to aquatic life and some links to Parkinson disease the compound is still allowed in organic farming as it is a naturally occurring compound.

Food Quality and Safety

While there may be some differences in the amounts of nutrients and anti-nutrients when organically produced food and conventionally produced food are compared, the variable nature of food production and handling makes it difficult to generalize results, and there is insufficient evidence to make claims that organic food is safer or healthier than conventional food. Claims that organic food tastes better are not supported by evidence.

Soil Conservation

Supporters claim that organically managed soil has a higher quality and higher water retention. This may help increase yields for organic farms in drought years. Organic farming can build up soil organic matter better than conventional no-till farming, which suggests long-term yield benefits from organic farming. An 18-year study of organic methods on nutrient-depleted soil concluded that conventional methods were superior for soil fertility and yield for nutrient-depleted soils in cold-temperate climates, arguing that much of the benefit from organic farming derives from imported materials that could not be regarded as self-sustaining.

In Dirt: The Erosion of Civilizations, geomorphologist David Montgomery outlines a coming crisis from soil erosion. Agriculture relies on roughly one meter of topsoil, and that is being depleted ten times faster than it is being replaced. No-till farming, which some claim depends upon pesticides, is one way to minimize erosion. However, a 2007 study by the USDA's Agricultural Research Service has found that manure applications in tilled organic farming are better at building up the soil than no-till.

Biodiversity

The conservation of natural resources and biodiversity is a core principle of organic production. Three broad management practices (prohibition/reduced use of chemical pesticides and inorganic fertilizers; sympathetic management of non-cropped habitats; and preservation of mixed farming) that are largely intrinsic (but not exclusive) to organic farming are particularly beneficial for farmland wildlife. Using practices that attract or introduce beneficial insects, provide habitat for birds and mammals, and provide conditions that increase soil biotic diversity serve to supply vital ecological services to organic production systems. Advantages to certified organic operations that implement these types of production practices include: 1) decreased dependence on outside fertility inputs; 2) reduced pest management costs; 3) more reliable sources of clean water; and 4) better pollination.

Nearly all non-crop, naturally occurring species observed in comparative farm land practice studies show a preference for organic farming both by abundance and diversity. An average of 30% more species inhabit organic farms. Birds, butterflies, soil microbes, beetles, earthworms, spiders, vegetation, and mammals are particularly affected. Lack of herbicides and pesticides improve biodiversity fitness and population density. Many weed species attract beneficial insects that improve soil qualities and forage on weed pests. Soil-bound organisms often benefit because of increased bacteria populations due to natural fertilizer such as manure, while experiencing reduced intake of herbicides and pesticides. Increased biodiversity, especially from beneficial soil microbes and mycorrhizae have been proposed as an explanation for the high yields experienced by some organic plots, especially in light of the differences seen in a 21-year comparison of organic and control fields.

Biodiversity from organic farming provides capital to humans. Species found in organic farms enhance sustainability by reducing human input (e.g., fertilizers, pesticides).

The USDA's Agricultural Marketing Service (AMS) published a *Federal Register* notice on 15 January 2016, announcing the National Organic Program (NOP) final guidance on Natural Resources and Biodiversity Conservation for Certified Organic Operations. Given the broad scope of natural resources which includes soil, water, wetland, woodland and wildlife, the guidance provides examples of practices that support the underlying conservation principles and demonstrate compliance with USDA organic regulations 205.200. The final guidance provides organic certifiers and farms with examples of production practices that support conservation principles and comply with the USDA organic regulations, which require operations to maintain or improve natural resources. The final guidance also clarifies the role of certified operations (to submit an OSP to a certifier), certifiers (ensure that the OSP describes or lists practices that explain the operator's monitoring plan and practices to support natural resources and biodiversity conservation), and inspectors (onsite inspection) in the implementation and verification of these production practices.

A wide range of organisms benefit from organic farming, but it is unclear whether organic methods confer greater benefits than conventional integrated agri-environmental programs. Organic farming is often presented as a more biodiversity-friendly practice, but the generality of the beneficial effects of organic farming is debated as the effects appear often species- and context-dependent, and current research has highlighted the need to quantify the relative effects of local- and landscape-scale management on farmland biodiversity. There are four key issues when comparing the impacts on biodiversity of organic and conventional farming: (1) It remains unclear whether a holistic whole-farm approach (i.e. organic) provides greater benefits to biodiversity than carefully targeted prescriptions applied to relatively small areas of cropped and non-cropped habitats within conventional agriculture (i.e. agri-environment schemes); (2) Many comparative studies encounter methodological problems, limiting their ability to draw quantitative conclusions; (3) Our knowledge of the impacts of organic farming in pastoral and upland agriculture is limited; (4) There remains a pressing need for longitudinal, system-level studies in order to address these issues and to fill in the gaps in our knowledge of the impacts of organic farming, before a full appraisal of its potential role in biodiversity conservation in agroecosystems can be made.

Organic Agriculture and the Environment

Ecosystem Approach in Organic Agriculture

In many countries, agriculture is the largest form of land use; farmland habitats account for a great percentage of natural habitats. Protected areas are not sufficient for nature conservation, especially for migratory species (e.g. birds) because surrounding habitats are agricultural lands which often create negative externalities. Agriculture, especially in its most extreme modern form of industrialized monoculture, modifies landscapes and damages the ecosystem's goods and services, including biodiversity at all levels. Agricultural encroachment, as well as pollution and intensification are responsible for contributing to soil and water degradation and to the extinction of biodiversity.

Current approaches to ecological agriculture such as integrated pest management, integrated plant nutrition systems and conservation tillage, consider only one aspect of the farming system components: pest ecology, plant ecology and soil ecology, respectively. The ability to define strategies that combine these and other management elements into a single approach is proper to organic agriculture. Organic management focuses on the food web relations and element cycling and aims to maximise the agro-ecosystem's stability and homeostasis.

Organic agriculture, through its systemic approach and avoidance of agro-chemicals, prevents natural resource degradation and the loss of land and productive potential. In organic agriculture, nature is both instrument and aim. As organic farmers cannot use synthetic substances (e.g. fertilizers, pesticides, pharmaceuticals) they need to restore the natural ecological balance because ecosystem functions are their main productive "input". For example:

- Many unspecific pests like aphids, thrips, whiteflies or spider mites, economically damaging in many crops, can be kept below the economic threshold with naturally occurring or purposely released predators and parasitoids. The former are direct goods and services of hedges, botanically diverse field margins, intercropping or weedy undergrowth, the latter do better when released in botanically and ecologically enriched habitats.

- The only way to suppress soil-borne pests and diseases in organic agriculture are wide crop rotations with several components of botanically different crops. Adhering to such rotations is crucial to providing agro-ecosystem diversity.

- Diverse crop rotations, or agro-forestry systems, guarantee a better uptake of nutrient elements from the soil and very efficient use of water and light, thanks to varying spatial and temporal root growth and leaf dispersion.

- Soils with a high functional diversity of micro-organisms, which occur very often after decades of organic agriculture practice, develop disease suppressive properties and can help to induce resistance in plants.

- By restricting farm inputs farmers have to use preventive techniques appropriately. The ban on herbicides, for instance, makes it impossible to ignore good crop rotation principles and disastrous in terms of yields and long-term problems with weeds. The ban on soluble and purchased fertilizers makes nutrient-conserving crop rotations and the sparing use of organic fertilizers to reduce losses, economically worthwhile.

As organic farmers cannot use synthetic substances (e.g. fertilizers, pesticides, pharmaceuticals) they need to restore the natural ecological balance because ecosystem functions are their main productive "input".

The direct economic value of ecosystem functions for organic farmers is an excellent example of "benefit-sharing". To master ecosystem functions in order to produce decent high-quality yields, adaptive management practices are crucial: growing conditions in organic agriculture are neither optimal, static nor predictable. Farmers must become excellent observers and must be trained to react flexibly, often intuitively and appropriately to the local context. In contrast, other farming methods try hard to condition the environment of the plants for optimal growing conditions by using a wide range of inputs. Moreover, most recommendations and fertilization and spraying programmes are highly standardized and not site-specific.

Organic agriculture is an approach, which is not confined to production alone. Intersectoral cooperation between farmers, environmentalists, processors and traders, quality controllers and consumers is a very common characteristic of organic agriculture.

Doherty et al. described sustainable agriculture as "ecological engineering solutions aiming to minimally manipulate and manage ecosystems for the benefit of both nature and humanity". To date, no ecological farming scheme other than the organic one has succeeded in providing ecosystem functions and socio-economic goods in a comparably equitable way.

Organic Agriculture and Abiotic Resources

Soil

As the central basis for all agricultural activity, soil is one of the chief natural resources. Since organic farmers cannot compensate for a loss in soil fertility by inputs of synthetic nutrients, the building and maintenance of soil fertility is a central objective of organic agriculture. The impact of organic agriculture on soil properties has been covered extensively by research with a special focus on the relevant parameters of organic matter content, biological activity and soil erosion.

Organic Matter Content

The environmental importance of organic matter content is its capacity to limit physical damage and to improve nutrient availability as well as biological activity. Research on organic matter concentrates on measuring the soil organic carbon content parameter.

The review of relevant research conducted by Stolze et al. concludes that under European conditions organic agriculture has beneficial effects on the characteristics of soil organic matter because the soil organic carbon content is higher on organically farmed soils than on conventional ones. Fertilization in organic agriculture is based on organic substances such as farmyard manure, compost, green manure, plant residues and commercial organic N-fertilizers. Consequently, there is an extensive supply of organic matter passing through aerobic decomposition processes.

Mineralization and decomposition processes are influenced by humidity, temperature and oxygen. Under humid tropical conditions these processes run faster and all year round, whereas under temperate conditions they are slower and come to a halt during the colder months. Soil type also plays a role. Sandy soils dry out quickly, slowing down the decomposition process, ferralitic soils on the other hand are generally not very fertile, but they encourage fast decomposition and the building-up of stable organic matter.

Biological Activity and Soil Micro-organisms

Intense biological activity promotes metabolism between soil and plants and should be a focus of sustainable plant production and fertilizer management. In contrast to conventional agriculture, organic farmers depend more on a high and sustained supply of organic substances including crop rotations with clover/grass ley, underseeds, catch crops, green and animal manure. Under European conditions, organic agriculture performs better than conventional agriculture with regard to certain parameters: for example, 30 to 100 percent higher microbial activity and a significantly higher biomass (+30 to 40 percent), density (+ 50 to 80 percent) and species diversity of earthworms, a key soil-macro faunal species.

Soil is a habitat for plants, animals and micro-organisms. As plants build up organic matter, soil animals feed on them and their debris, whilst microbes decompose the complex organic compounds to their mineral components and to carbon dioxide. A living soil is central to soil fertility because it is the activity of soil organisms that makes available the elements in plant residues and organic debris entering the soil. Part of this material, however, remains in the soil and contributes to its stabilization by humus build-up.

Various research results show that the activity of micro-organisms is higher in organically- than in conventionally-managed soil. As a consequence, in organically managed soils, nutrients are recycled faster and soil structure is improved. For example, found in Switzerland up to 90 percent higher total mass of micro-organisms in organically managed soils. As for soil fungi, found a higher number and abundance of saprophytic soil fungi with a higher potential of decomposition of organic material. Mycorrhizae, important representatives of soil fungi, live as symbionts with plant roots. The degree of mycorrhizal root colonization was found to be distinctly higher in organic plots as compared to conventional plots.

As mineralization processes proceed much faster on ferralitic soils in the tropics and subtropics than on soils typical to temperate and continental zones, a high organic matter content and high biological activity are the pre-requisite for sustainable soil fertility. The positive impacts of organic agriculture on biological activity, micro-organisms and soil organic matter content reported in the paragraphs above are therefore also valid for soils in the tropics and subtropics.

Soil Erosion

Soil erosion is assumed to be the main cause of soil degradation around the world. The loss of fertile top soil by erosion results in a lower yield capacity on the onehand and in a undesired transfer of nutrients, pesticides and sediments in surface water on the other.

Reganold et al. found during a long-term comparison of soils from selected conventional and organic farms of the same soil type near Spokane in Washington, USA, that topsoil was 16 cm thicker on organically managed plots. This was most probably due to inclusion of a green manure legume crop in the third year of rotation and fewer tillage operations on the organic field. The organically-farmed soil not only had deeper topsoil but also significantly higher organic matter content and suffered less soil erosion than the conventionally farmed soil. The authors concluded that the organic agriculture system was more effective than the conventional agriculture system in reducing soil erosion and in maintaining soil productivity.

Generally, organic soil management techniques such as organic fertilization, mulching and cover cropping improve soil structure and therefore increase the soil's water infiltration and retention capacity, substantially reducing the risk of erosion. These management techniques are also of special relevance on porous ferralitic soils of the tropics and subtropics which are highly susceptible to soil erosion due to frequent heavy rainfall.

Synthesis of Research Results on Soil and Organic Agriculture

A thorough comparison of relevant soil parameters on conventionally- and organically-managed soils is provided by the long-term DOC-trial carried out by the Swiss Research Institute of Organic Farming (FiBL). The results of the DOC trial can be summarized as follows:

- Organic matter content is usually higher in organically-managed soils than in exclusively minerally-fertilized conventionally-managed ones, thanks to organic fertilization methods. High organic matter content also helps to avoid soil acidification.
- Organic soil management improves soil structure by increasing soil activity, thus reducing the risk of erosion.
- Organic crops profit from root symbioses and are better able to exploit the soil.
- Organically farmed soils have significantly higher biological activity (earthworms, fungi, bacteria, micro-organisms) than conventionally-managed ones. Thus nutrients are transformed more rapidly and soil structure is improved.
- Organic management promotes the development of soil fauna such as earthworms and above ground arthropods, improving the growth conditions of the crop. More abundant predators help to control harmful organisms.

Organic soil management improves soil structure by increasing soil activity, thus reducing the risk of erosion.

Ground and Surface Water

The detrimental effects of intensive agriculture on ground and surface water are largely due to erosion and to nitrate and pesticide pollution. The chief threats to water quality posed by agriculture are: high organic fertilization levels in combination with high stocking rates, the excessive application of mineral N-fertilizers; the lack of a protective soil cover; narrow crop rotations and frequent tillage; high levels of available nitrogen after harvest, and contamination of water with synthetic pesticides. As organic agriculture uses no synthetic pesticides, there is no risk of synthetic pesticide pollution of ground and surface waters.

The reasons for the lower nitrate leaching rates in organic agriculture are the ban on mineral N-fertilizers and lower livestock densities. These constraints set up by the organic agriculture standards lead to nitrogen on organic farms nitrogen being - in economic terms - a quantitatively scarce factor. The economic consequences of nitrogen being scarce on organic farms are quite impressive: the opportunity costs (cost of producing on-farm) of 1 kg nitrogen on organic farms can amount to seven to sixteen times the cost of mineral N-fertilizers. So it is not surprising that in contrast to conventional farms where manure and slurry are often a waste problem, organic farmers are forced to develop efficient nitrogen management strategies like intercropping, catch cropping, optimal ploughing-in of legume crops or limiting the use of liquid manure to avoid nitrogen losses.

As a result, in some regions of Germany and France, waterworks subsidise conversion to organic agriculture in water protection areas as an economically efficient solution for reducing the cost of cleaning-up drinking water by minimising the nitrate and pesticide contamination of groundwater.

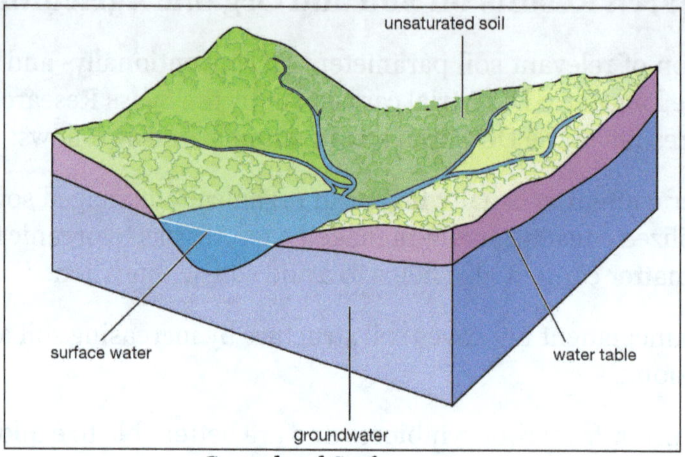

Ground and Surface water.

Waterworks subsidise conversion to organic agriculture in water protection areas as an economically efficient solution for reducing the cost of cleaning-up drinking water by minimising the nitrate and pesticide contamination of groundwater.

While scientific results from other climatic zones are scarce, positive effects of organic agriculture on the nitrate leaching risk can be reported from a citrus farm in Cuba. Under organic fertilization management based on compost with 60 kg available N per ha, the farm achieved exactly the same yield level as under conventional fertilization management with 200 kg of mineral N. This example shows that organic fertilization management can help reduce the risk of nitrate leaching especially under extreme climatic conditions.

Nutrient Use

An adequate and balanced supply of nutrients in the soil is essential for several reasons. Nutrient surpluses might result in nutrient losses which could subsequently lead to water and air contamination and eutrophication. However, nutrient deficiency is synonymous with the over-exploitation of soil nutrients in the long run, which consequently leads to a decrease in yield and crop quality.

Research by Freyer in Switzerland shows that only 14 percent of all organic farms have an N-surplus, and only 1.5 percent had a P-surplus. Most of the organic farms have a negative N- and P-balance.

Table: Examples for P, K balances (kg/ha) comparing organic with conventional farms from different European countries.

	P balance (kg/ha)		K balance (kg/ha)	
	Organic	Conventional	Organic	Conventional
Sweden	-12	+37	-4	+39
Netherlands				
Cash crop farm	+18	+23	+31	+25

Horticulture	+32	+60	+119	+110
Dairy farm	+8	+31	-	-
Germany				
Mixed farm	-4	+13	-27	+31
Dairy farm	-2	+5	+7	+20

Due to negative nutrient balances the question arises whether organic agriculture methods cause gradual loss of soil minerals. First of all, the proportion of soluble nutrients is lower on organically-managed soils. On the other hand, found no decrease in organic yields as an indicator for nutrient deficiency on farms which are managed organically for more than 30 years. The higher biological activity and higher mycorrhizal root colonization counteract nutrient deficiency, thus as state, for phosphorus, the aim of organic agriculture of increasing nutrient supply through increased biological activity has been achieved. Higher biological activity and higher mycorrhizal root colonization counteract nutrient deficiency.

Energy Use

Energy consumption in agriculture includes the direct consumption of fossil energy (e.g. fuel and oil), as well as indirect energy consumption (e.g. from the production of synthetic fertilizers and pesticides). Leaving aside indirect agricultural energy consumption, OECD statistics indicate that agriculture contributes only 2 percent to total direct energy used in OECD countries. Nevertheless, limited fossil energy resources and the climatic relevance of its use require efficient energy use, even in agriculture. Relevant parameters for evaluating energy use in agriculture are energy consumption and energy efficiency.

Considering both, direct and indirect energy consumption, scientific calculations on energy consumption per hectare indicate that organic farms use less energy than conventional farms: several researchers calculated the energy consumption of organic farms to amount 64 percent of that on conventional farms. Other recent research confirms the figures mentioned above at lower levels, with energy consumption on organic farms amounting to 45 percent or 30 to 50 percent of conventional farms, respectively.

The determining factor for energy consumption of a specific crop is cropping management, which includes tillage intensity, manuring and weed control. On a per hectare scale, all authors determined lower energy consumption on organic farms. For organic potatoes and apples, however, energy consumption per output unit is higher relative to conventional production. This is the result of higher energy input for mechanical measures like weed control in organic production and of lower mineral N-fertilization use in conventional production.

A second parameter appropriate for evaluating energy use is energy efficiency. It provides information about the ratio of energy input and output. Comparing rotations of different production systems in Iran, found the energy efficiency of organic agriculture to be 81 percent better compared to high-input conventional agriculture. In a similar investigation in Poland Kus and Stalenga calculated a 35 percent higher energy efficiency of organic compared to conventional agriculture. Under Mediterranean conditions in Italy, a 25 percent higher efficiency in organic wheat and an 81 percent higher efficiency in organic vineyard production systems were found.

Even though the ban on synthetic pesticides might lead to higher fuel consumption on organic farms due to increased mechanical weed control, research results presented below show that with respect to energy consumption, organic agriculture is performing better than conventional agriculture. The main reasons for this are:

- No input of mineral N-fertilizers which require a high energy consumption for production and transport on organic farms;
- Lower use of highly energy-consumptive foodstuffs (concentrates);
- Lower input of mineral fertilizers (P, K);
- Ban on synthetic pesticides.

With respect to energy consumption, organic agriculture is performing better than conventional agriculture.

Table: Calculations of energy consumption of different products.

Product	Energy use GJ/ha			Energy use GJ/t		
	Conventional	Organic	as % of conventional	Conventional	Organic	as % of conventional
Winter wheat						
Alföldi et al.	18.3	10.8	-41	4.21	2.84	-33
Haas and Köpke	17.2	6.1	-65	2.70	1.52	-43
Reitmayr	16.5	8.2	-51	2.38	1.89	-21
Potatoes						
Alföldi et al.	38.2	27.5	-28	0.07	0.08	+7
Haas and Köpke	24.0	13.1	-46	0.08	0.07	-18
Reitmayr	19.7	14.3	-27	0.05	0.07	+29
Citrus						
Barbera and La Mantia	43.3	24.9	-43	1.24	0.83	-33
Olive						
Barbera and La Mantia	23.8	10.4	-56	23.8	13.0	-45
Apple						
Geier et al.	37.35	33.8	-9.5	1.73	2.13	+23
Milk						
Cederberg and Mattsson	22.2	17.2	-23	2.85	2.41	-15
Wetterich and Haas	19.1	5.9	-69	2.65	1.21	-54

However, higher labour requirements may be expected on organic farms under European conditions due to a higher share in the production of labour intensive crops (e.g. vegetables) and on-farm marketing and processing activities. Arable crops also demand greater labour inputs, for example, for mechanical weeding.

In Europe, figures for labour use on organic farms in relation to comparable conventional farms vary between countries and studies. Most commonly, labour use per hectare of utilized agricultural area is on average 10-20 percent higher on organic farms. In a European context, the labour input is higher on organic arable and mixed farms, while organic dairy farms use the same amount of labour, or less, than comparable conventional farms. On horticultural farms, labour requirements are much higher than on conventional farms. Few data exist on pig and poultry farms, but labour per hectare of utilized agricultural area seems to be similar to conventional farms, as livestock density is reduced.

Organic Agriculture and Biodiversity

Agriculture and Biodiversity

For hundreds of years, agriculture has contributed substantially to the diversity of species and habitats, and agriculture has formed many of today's landscapes. Over the last century, however, modern intensive agriculture, with its high input of synthetic pesticides and fertilizers and monocrop specialization, has been detrimental to the diversity of genetic resources of crop varieties and livestock breeds, to the diversity of wild flora and fauna species and to the diversity of ecosystems. The 2000 IUCN Red List of threatened species highlights habitat loss as the main threat to biodiversity, with agricultural activities affecting 70 percent of all threatened bird species and 49 percent of all plant species.

Because of the increasing loss of agricultural biodiversity on a global scale, the Convention on Biological Diversity has developed a work programme on this subject in 1996. This programme states inter alia that farming practices that stop degradation while restoring and enhancing biological diversity should be encouraged, including organic agriculture.

Organic agriculture is dependent upon stabilizing agro-ecosystems, maintaining ecological balances, developing biological processes to their optimum, and linking agricultural activities with the conservation of biodiversity. Wild species perform a variety of ecological services within organic systems: pollination, pest control, maintenance of soil fertility. Thus, higher levels of biodiversity can strengthen functions essential for farming systems and therefore, agricultural performance. Enhancing functional biodiversity is a key ecological strategy to bring sustainability to production on organic farms. Organic systems also use substantially fewer external inputs and do not use synthetic chemical fertilizers, pesticides, genetic modified organisms and pharmaceuticals. Instead, systems are designed to work in harmony with nature in order to determine agricultural yields and disease resistance. By respecting the natural capacity of plants, animals and the landscape, organic agriculture aims to optimize quality in all aspects of agriculture and the environment.

Biological pest control on organic farms, for example, relies on maintaining healthy populations of pest predators and parasitoids. A study in California comparing conventional with organic tomato fields showed a higher abundance of natural enemies and greater richness of species in organic tomato fields. There was no significant difference for any type of damage to tomato foliage or fruit, showing that the organic system achieves the same levels of pest control without having to apply synthetic chemical pesticides.

Organic agriculture is thus committed to the conservation and enhancement of biodiversity within agricultural systems, both from a philosophical perspective and from the practical viewpoint of maintaining productivity. To this end, the importance of biodiversity as part of a well-balanced organic system is part of the IFOAM International Basic Standards for Organic Agriculture Production and Processing.

Numerous scientific studies, mainly from Europe and North America, give evidence that on organic farms biodiversity is higher than that on conventional and farms. Biodiversity is generally assessed at three distinct levels:

- Genetic diversity: The variation between individuals and between populations within a species;
- Species diversity: The different types of plants, animals and other life forms within a region or community;
- Ecosystem diversity: The variety of habitats found within an area (grassland, marsh, and woodland for instance).

These levels of biodiversity have been used in the following topics to show the links between biodiversity and organic agriculture.

In general, the degree of biodiversity in agro-ecosystems depends on four main characteristics of agro-ecosystems:

- The diversity of vegetation within and around the agro-ecosystem.
- The permanence of various crops within the agro-ecosystem.
- The intensity of management/farming.
- The extent of isolation of the agro-ecosystem from natural vegetation.

Diversity of Genetic Resources in Organic Agriculture

Higher Crop Diversity on Organic Farms

Since organic farms are mostly mixed farms, integrating animal husbandry with crop production, using vast and diverse rotations, intercrops and green cover crops, and maintaining soil fertility by cultivating nitrogen fixing legumes, they display a higher diversity of domesticated species than conventional farms.

Hausheer et al. evaluated crop rotations on 110 organic, integrated and conventional farms in a Swiss pilot farm project. They found more diverse crop rotations (4.5 different crops in organic as opposed to 3.4 different crops in integrated farming) and a higher number of crops, including perennials, vegetables and herbs (10.2 in organic and integrated farms; 7.4 in conventional farms).

Maintenance of Genetic Resources in Organic Agriculture

Today, the adoption of high-yielding, uniform cultivars and varieties has led to a considerable reduction in the number of plants and animals used in agriculture. Only 120 cultivated plant species and 14 mammalian and avian species provide 90 percent of human food supply.

Evidence of the trend towards monoculture and uniformity is given by the fact that in India, under the Green Revolution, the number of cultivated rice varieties has decreased from more than 100 000 to 10; also, 50 percent of the goat breeds, 20 percent of the cattle breeds and 30 percent of the sheep breeds are in danger of extinction. In Mexico, only 20 percent of the maize varieties reported in 1930 are now known. In China, nearly 10 000 wheat varieties were used in production in 1949; by the 1970s, only about 1000 remained in use. This trend is just as visible in animals: 740 animal breeds became extinct during the twentieth century. Currently, 1350 breeds face extinction, with two breeds being lost each week.

Organic producers look for productive varieties suited to their local climatic and soil conditions and that are not susceptible to disease and pest attack. Organic agriculture standards recommend the cultivation of site adapted crop varieties, characteristics often found in the older native cultivars. This, however, does not necessarily mean that organic agriculture sets narrow limits to the use of modern maximum yield varieties, which are often chosen for pest/disease resistance purposes. Still, the preservation of native varieties and breeds is an important initiative of the organic movement, but their actual use depends on individual farmers.

Many seed banks and indigenous variety conservation programmes worldwide are linked to organic agriculture projects. For instance, the Sustainable Agriculture and Rural Development Project in Kenya is working with communities in the Gilgil district to develop organic systems to increase food security through a community indigenous seed conservation programme. Indigenous seeds have been shown to perform better in the harsh drought conditions.

Genetic Engineering, Agricultural Biodiversity and Organic Agriculture

Apart from the adoption of high-yielding, uniform cultivars, a further possible threat to genetic diversity, and biodiversity in general, are the side-effects of the release of genetically engineered organisms into the environment.

Genetically engineered plants designed to control pests can have negative side-effects on beneficial insects and further non-target organisms as well. Oilseed rape with genetically induced resistance to insects has been reported to damage beneficial insects such as honey bees. The use of herbicide resistant plants can result in greater use of herbicides, increasing the negative effects of intensive farming on natural biodiversity. Furthermore there is the danger that transgenic plants could become feral and thus suppress indigenous flora. Feral oilseed rape populations in Canada are resistant to three herbicides and have become one of the most troublesome weeds.

Recently, it was reported that transgenic DNA introgressed into traditional maize landraces in Oaxaca, Mexico. Concerns have been raised about the potential effects of transgenic introductions on the genetic diversity of crop landraces and wild relatives in areas of crop origin and diversification, as this diversity is considered essential for global food security.

Insofar as organic agriculture is dependent upon maintaining ecological balances and diverse agro-ecosystems, genetic engineering is a contradiction to the principal aims of organic agriculture. Organic agriculture does not allow genetic engineering in its standards as genetic engineering focuses on genetic makeup without taking into account the complete organism or farming system in which the organism functions.

Species Diversity and Organic Agriculture

Floral diversity has been shown to contribute to ecosystem stability, while the invertebrate community associated with field boundaries performs many ecosystem functions including biological control of pests and diseases, pollination and food resource for higher trophic levels.

Floral Diversity

Today, the diversity of the typical wild flora on arable fields, which is the main habitat for a wide range of species, is at risk. Many species are endangered due to agricultural intensification, including the intensive use of mineral fertilizers and of herbicides, intensive soil management and the destruction of habitats. In grassland species, diversity is also decreasing due to intensification of grazing management and to high inputs of fertilizers.

Whereas in conventional agriculture weeds are considered competitive to the crop and are eliminated by herbicides, in organic systems some of the accompanying plants are desired to a certain degree and considered useful, as they provide a wide range of ecological services. These services include protection from soil erosion, providing shelter and alternative food resource for beneficial organisms and pollinators.

Figure shows the results of several studies comparing floral diversity on organic and on conventional fields. In all cases, organic agriculture showed higher floral diversity. Several comparative analyses carried out in Europe have shown plant species diversity in organic arable fields to be 30 to 350 percent higher than in conventional ones. In field edges, the number of plant species can be twice as high as in conventional fields, and the inner parts of the fields can have up to six times as many species compared to conventional fields.

In Sweden, a number of declining, endangered and rare species were recorded on organic arable fields, showing that organic agriculture can contribute to maintaining biodiversity.

In Romania, a weed survey in maize fields showed a significantly higher diversity of weed species in the organic maize field. This study had been carried out within the framework of a research project assessing the feasibility of organic agriculture in three Danube river countries. In this study positive effects of the organic system had also been recorded for further biodiversity indicators (e.g. ground-dwelling arthropods and earthworms).

In most cases, organic agriculture conserves better than conventional agriculture the site-typical plant community of floral species of arable land. A survey comparing organic with conventional fields showed that in the organic system the share of the fields with endangered floral species was 79 percent as opposed to 81 percent 27 years earlier, showing that the share had almost not changed. In the conventional fields the rate had dropped from 61 percent to 29 percent.

The higher floral diversity and abundance in organic arable fields is generally due to the ban on synthetic N-fertilizers and herbicides. The limited availability and input of nitrogen, the application of mechanical and thermal weed control and more diverse crop rotations and a higher crop diversity lead to more favourable conditions for many wild plant species.

In organic grassland, the average number of species was found to be around 25 percent higher in organic than in conventional grassland, including some species in decline. The plant community structure in organic grassland is more even and more typical for a specific site than in conventional agriculture. The higher floral diversity is often caused by lower stocking rates and lower fertilization levels in organic farms as well as to later mowing dates, especially in meadows, which means that grass species can reach the flowering stage and thus have a higher reproduction rate.

Floral Diversity.

The limited availability and input of nitrogen, the application of mechanical and thermal weed control and more diverse crop rotations and a higher crop diversity lead to more favourable conditions for many wild plant species.

Effects of Vegetation on Faunal Diversity

Weeds influence the diversity and abundance of arthropods (e.g. beetles, ants and spiders), acting as natural food resource and shelter. Weeds like Umbelliferae, Leguminosae and Compositae play an especially important ecological role as they provide food and thus improve reproduction of many arthropod species. Research carried out on tomato plots on the effects of weed control on surface-dwelling arthropod species found the abundance of species is clearly influenced by weed biomass. Species numbers were lowest where mulching with rye straw was controlling the weeds. However, removing weeds within 20 cm of each plant reduced weed biomass but retained higher arthropod populations than in the plots treated with herbicide or mulch.

With regard to pollinators, which greatly benefit from a richness of flowers, the fact that flowering weeds are more diverse and more abundant in organic arable fields and in organic grassland compared to conventional fields, where only few species and numbers were found, is particularly important. Many insect species which feed on nectar and pollen have a higher reproduction rate when they benefit from a better food supply through species-rich plant communities.

It may be assumed that, as for butterflies, the organic system also favours an abundance of further pollinators like bees and wasps. Flowering plants are also important for many beneficial arthropods such as predators and parasitoids.

Orchards with rich floral undergrowth have a lower incidence of insect pests than orchards treated with herbicides, mainly because of an increased abundance and efficiency of predators and parasitoids. Similarly, in other permanent crops, cover crops do not only provide erosion control and nutrient supply but also a high floral and faunal species diversity. This is reported for organic olive production and for organic vine growing, for instance in California.

In order to diversify the farming system and attract beneficial arthropods and pollinators, wild flower strips are sown in organic agriculture orchards. In a Swiss organic orchard, it was found that the strip management favoured beneficial insects and spiders, which reduced the density of aphids. The density of aphids was reduced due to higher mortality caused by increased numbers of predators feeding on aphids. Measures aimed at managing appropriate habitats and thus increasing floral and structural diversity is a key strategy for improved natural pest control.

Faunal Diversity

Organic agriculture displays in most cases a higher faunal biodiversity than conventional agriculture. Apart from the better food resources in organic fields, the key factors are more fauna-compatible plant protection management, organic fertilization, the more diversified crop rotation and the more structured landscapes with semi-natural habitats and field margins.

The effects of organic agriculture on faunal biodiversity have been studied particularly for soil fauna and for birds. A review of 44 research studies in Europe and the United States of America (on farm and pilot trials) on the effects of farming systems on beneficial invertebrates and birds consistently shows a better performance of the organic system. Of the faunal groups analysed (i.e. earthworms, arthropods and birds) in 49 out of 55 investigations organic agriculture performed better in terms of abundance. As regards species diversity, in 15 out of 23 investigations organic agriculture performed better; in no case conventional farming had a better performance.

Research studies in Europe and the United States of America (on farm and pilot trials) on the effects of farming systems on beneficial invertebrates and birds consistently shows a better performance of the organic system.

The cases where biodiversity in organic systems is not better than conventional systems might be explained by the presence of robust species which do not have special requirements regarding their habitats, such as some spider species.

Earthworms

Earthworms are highly suitable bio-indicators of soil fertility, and they are known for their sensitivity to synthetic pesticides and to many agricultural practices. Due to their biology, earthworm populations can indicate the structural, microclimatic, nutritive and toxic situation in soils. In conventional agriculture, earthworms are affected by the use of harmful pesticides and intensive soil cultivation.

Earthworms generally increase nutrient cycling rates. Their casts greatly help to improve soil structure and have high concentration of nutrients in an accessible form to plants. The burrowing activity of earthworms enhances aeration, porosity and drainage of the soil, all of which are important factors in the development of healthy and well-developed crop root system. Earthworms also play an important role in pest and disease control, including the reduction of leaf miner pupae and scab pathogens in orchards.

Many investigations in Europe and North America provide evidence that generally, organically managed soils exhibit a higher abundance and species number than conventionally-managed plots or farms. The biomass of earthworms in the organic system in the Swiss DOC long-term trial was 30 to 40 percent higher than in the conventional system, the number of individuals even 50 to 80 percent higher. Vertically-burrowing earthworm species which are of high agro-ecological relevance (e.g. for water infiltration, soil aeration, reduction of erosion) were enhanced by organic agriculture. Reasons for the higher presence of these species in the biological plots might be species-specific effects of pesticides. Vertically burrowing species are more exposed to pesticides because they are crawling on the ground surface for feeding and mating. Furthermore they can be harmed by pesticide contaminated water flowing into their stable burrows. The decrease in juvenile earthworms in the conventional plots can be explained by the fact that upper soil layers are their preferred habitat. Thus, they may be exposed to higher pesticide concentrations than adults, because juveniles are unable to escape into deeper soil layers. In an organic barley field in Bulgaria in 1998, a single cubic meter of soil contained 124 earthworms compared to only 21 in conventional soil.

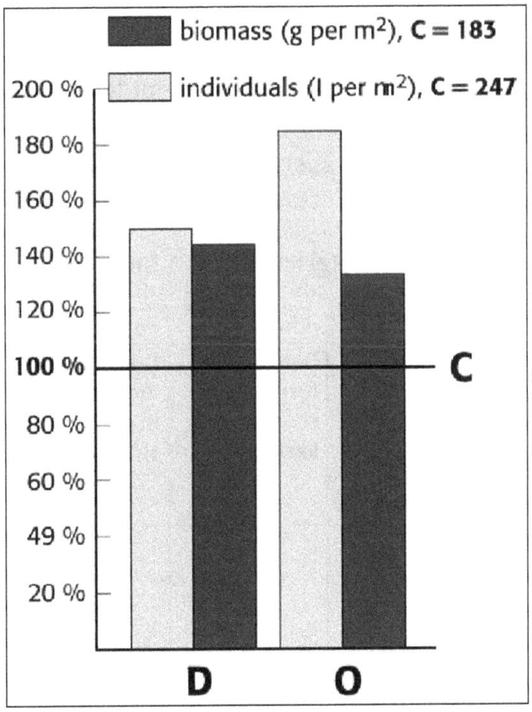

Biomass and density of earthworms in the DOC long-term trial.

Organic matter constitutes an important food source for earthworms and can be maintained by an appropriate fertilizing and crop rotation system. Investigations have shown that earthworms also benefit greatly from green manuring and the planting of grass-clover in the crop rotation. The less intensive farming practices (plant protection, soil cultivation), the more favourable fertilizing management and the more diversified crop rotations on organic farms enhance specimen- and species-rich earthworm populations.

Arthropods

Beneficial arthropods that live above the ground include ground beetles (carabids), rove beetles (staphilinids) and spiders. Many of these polyphagous species, which feed on a wide range of food items, are important predators, and in arable crops they play an important role in the regulation of various pests. In addition, certain arthropods, especially carabids, are considered as sensitive indicators of habitat quality. Monitoring them yields useful information on the sustainability of different agricultural farming systems.

In conventional agriculture, synthetic pesticides can have negative impacts on beneficial arthropods. Pesticides affect arthropods either directly, via contamination or reduction of their prey, or through alterations of the micro-habitat. The reproduction rate may be reduced by sublethal long-term effects of synthetic pesticides used in conventional agriculture. Foliar fungicides can lead to mortality in springtails and thus negatively influence polyphagous predators. Higher fertilization in conventional fields leads to higher crop density, and this in turn can alter the micro-climate as well as reducing the occurrence of species dependent on a warm micro-climate and on light. This phenomenon was found in the DOC long-term plot-trial as well as on farm level. Organic fertilization can enhance epigeic arthropods through a richer supply of mesofauna decomposing organic compounds.

In several investigations of on-farm sites as well as in plot trials, a higher diversity and abundance of arthropods was found in organic and biodynamic plots compared to conventional plots. In the DOC long-term trial, the organic and biodynamic treatment had up to 100 percent more carabids, 60 to 70 percent more staphylinids and 70 to 120 percent more spiders. The organic fields were also characterized by a generally more even distribution of species, which means that the community is less dominated by few species only.

Table: Effects of organic and conventional agriculture on fauna.

Animal group	Abundance – number of individuals			Species diversity – number of species		
	Better performance in orgenic agriculture	No significant difference between orgenic and conventional agriculture	Better performance in conventional agriculture	Better performance in orgenic agriculture	No significant difference between orgenic and conventional agriculture	Better performance in conventional agriculture
Earthworm	17	1	0	4	3	0
Arthropods						
Carabids	13	2	0	6	2	0
Spiders	6	1	0	0	0	0
Myriapods	4	0	0	1	1	0
Bugs	2	1	0	1	1	0
Mites	2	0	1	1	1	0
Birds	5	0	0	2	0	0
Totals cases of al animals groups	49	5	1	15	8	0

In several investigations of on-farm sites as well as in plot trials, a higher diversity and abundance of arthropods was found in organic and biodynamic plots compared to conventional plots.

Researchers have also found greater diversity and abundance of many other invertebrate species in organic agriculture systems. For example, a study comparing spider communities in organic and conventional winter wheat fields in the United Kingdom found the abundance and diversity of spiders greater on the organic fields. It was concluded that the results were highly affected by the increased levels of under storey vegetation (i.e. broad-leaved and grass species) in the organic fields. Significantly more non-pest butterflies were recorded on organic farms than conventional farms but there was no significant difference in the abundance of pest species.

The quality and amount of food are key factors for the survival of the arthropod populations, which find a higher proportion of suitable food resources in organic fields including flowering non-grass plants.

Diversity of arthropods.

Various typical agricultural practices (e.g. ban of synthetic pesticides, organic fertilization, and habitat diversity) on organic farms are less detrimental for arthropod species compared with conventional agriculture.

Birds

Birds are well-suited indicator organisms showing the environmental status of nature and landscape infrastructure including agricultural land. Many bird species feed on insects, and an abundant presence may thus also contribute to a better natural pest control.

The breeding populations of a number of bird species in Western European farmland have declined in the past decades. This development is probably linked to intensified agricultural practices, including a general reduction in crop diversity, an increase in average field size, simplified rotations, cultivation of natural habitats, drainage, and increasing use of fertilizers and synthetic pesticides. Investigations have found a relationship between pesticide use and a decrease in quality and amount of food available for birds on conventionally farmed land. A consequence is a decreased breeding success of birds.

Several studies show that bird densities are higher in organic farms. A study by Rhône-Poulenc has shown a steady annual increase in the number of bird territories on land converted to organic production and a higher overall number of territories on the organically managed land.

A study by the British Trust for Ornithology compared breeding and over-wintering of birds on 44 organic and conventional farms. The study concluded that breeding densities of skylarks were significantly higher on organic farms and generally higher densities of birds, especially in winter, were found on the organic farms.

A three-year study in Denmark concentrated on the non-crop habitats, such as hedgerows, of conventional and organic farms and their effects on bird populations. The abundance of birds was 2 - 2.7 times greater on the organic farms. In total, 24 species where more prevalent on organic as opposed to conventional farms, of these 11 species had declined in number in Denmark since 1976.

Christensen et al. investigated a total of 31 farms comparing the bird densities on organic land with those on conventionally managed land. On one of the sites investigated they found a mean population density of 22.8 pairs of breeding birds per ten hectares on the organic land compared to 9.9 pairs on the conventional land. Also flora and invertebrate fauna in organically versus conventionally managed land had been compared on the farms selected for the bird survey. By late June, the organic fields had 50 to 70 percent more species of wild plants than the conventional fields, with significantly more weeds and biomass. These differences increased markedly during the growth season due to herbicide treatment in the conventional areas. The average biomass in the organic fields was significantly more stable than in conventional fields, resulting in greater security of food supply. These results suggest that limited food availability is a key factor in the reduced number of birds in conventionally farmed areas.

In the coffee producing countries of Latin America, growing organic coffee and cocoa under shade can have a major impact on biodiversity. Research carried out by the Smithsonian Migratory Bird Centre in Colombia and Mexico showed over 90 percent fewer bird species in sun-grown coffee plantations as opposed to shade-grown coffee. Although organic standards do not explicitly state the need for coffee to be grown in the shade, shade-grown practice is recommended as it fulfils re-quire-ments to enhance soil fertility, pest and disease control and expands crop production option.

The reasons for the higher number and species diversity of birds and bird territories on organic farms include better breeding habitats and better food conditions. Further reasons for the higher bird species diversity and abundance are the absence of synthetic pesticides, more semi-natural habitats (hedges, field margins) and higher crop diversity on organic farms.

Other Animals

Little work is available on the effects of organic agriculture on other wild animal groups. This could, however, be an important research theme, particularly with respect to pollinators. Many plants require pollen from other individuals to seed and regenerate.

Reasons for the higher number and species diversity of birds and bird territories on organic farms include better breeding habitats and better food conditions.

Over 80 percent of all flowering plant species and over three-quarters of the major world crops rely on animal pollinators. The principle pollinators are bees: approximately 73 percent of the world's cultivated crops, such as cashews, squash, mangoes, cocoa, cranberries and blueberries, are pollinated by some variety of bees, 19 percent of flies, 6.5 percent by bats, 5 percent by wasps, 5 percent by beetles, 4 percent by birds, and 4 percent by butterflies and moths. Of the hundred principal crops that make up most of the world's food supply, only 15 percent are pollinated by domestic bees (mostly honey bees, bumble bees and alfalfa leafcutter bees), while at least 80 percent are pollinated by wild bees and other wildlife (as there are an estimated 25 000 bee species, the total number of pollinators probably exceeds 40 000 species). Services of native pollinators are estimated to be worth US $4.1 billion a year to United States agriculture alone. Pollinators provide an essential ecosystem service that contributes to the maintenance of biodiversity, and ensures the survival of plant species including plants that provide food security to numerous households.

Declines in populations of pollinators now threaten both the yields of major food crops and the survival of wild plant species. Due to an epidemic of mites, a quarter of North America's wild and domestic honeybees have disappeared since 1988, with a cost to American farmers of US $5.7 billion per year.

Currently, 82 species of mammalian pollinators, including bats, 103 species of avian pollinators and one reptile are considered threatened or extinct according to IUCN criteria, the ratio of threatened vertebrate pollinators to the total numbers of vertebrates in their genera being extremely high. Geographically, large numbers of vertebrate pollinators are at risk in Australia, Colombia, Ecuador, Indonesia, Madagascar, Mexico, Papua New Guinea, Peru, and the United States. Common threats include: loss of nesting and roosting sites, habitat fragmentation by conversion or destruction of vegetation, habitat fragmentation by excessive exposure of nectar plants to herbicides and pollinators to pesticides, over-hunting, disruption of nectar corridors required by migratory pollinators, and competition by invasive species.

By not using synthetic chemical pesticides or herbicides and enhancing ecosystem diversity, it may be assumed that organic agriculture benefits pollinators and contributes to their conservation and survival.

Ecosystem Diversity

According to the World Resources Institute, an ecosystem is made up of the organisms of a particular habitat, such as a farm or forest, together with the physical landscape in which they live. Although little research has been carried out comparing agro-ecosystem diversity in different farming regimes, many of the principles of organic agriculture are likely to have a positive impact on ecosystem diversity.

Ecosystem Diversity in the Standards for Organic Agriculture

The 2002, IFOAM Basic Standards for Organic Agriculture Production and Processing include principles and recommendations on "organic ecosystems" where provisions are made to "maintain a significant portion of farms to facilitate biodiversity and nature conservation", including (among others) wildlife refuge habitats and wildlife corridors that provide linkages and connectivity to native habitat.

The Swiss organic standards require organic farmers to use 7 percent of their land as semi-natural habitats, including field margins. Many other organic certification bodies have included biodiversity requirements in their standards. This applies to organic and low-input farmers receiving direct payments under the agri-environment schemes. Direct payments by the state are linked to ecological performance which is controlled by inspection services.

In Sweden, a working group made up of organic farmer groups, nature conservationists, Government agencies and universities has been working since 1997 to strengthen the links between organic agriculture and biodiversity conservation, and they set up biodiversity plans for organic farms. The main objectives, apart from helping organic agriculture enhance biodiversity, have been to start cooperation and dialogue between the nature conservation and organic agricultural movements and to spread knowledge about biodiversity in organic agriculture. From there, the discussions developed into a planning exercise leading to proposals for changes to the country's main organic standard (set by the certification body, KRAV), to require all organic farmers to have a plan for the management of biodiversity on their farms from 2001.

Organic agriculture has a high and possibly decisive potential for reversing the dramatic decline of biodiversity. However, further efforts must be taken to refine the basic organic standards for biodiversity and landscape. Actual experiences show that there is a considerable and diverse synergy potential between biodiversity conservation and farm income. Biodiversity issues based on measurable performance are more and more important as a basis for subsidies and, at least in some countries, for consumer decisions for buying organic products.

Semi-natural Habitats as Part of Organic Agriculture

The conservation and the management of semi-natural habitats play an important role in organic agriculture. Semi-natural habitats are refuges for endangered plant species which in former times were found in meadows and arable fields. Semi-natural habitats and field margins are also essential for the survival of many invertebrates, especially due to favourable food and over wintering conditions. They also function as habitat cross-links between meadows, fallows and different field margins.

Quantitative data from Switzerland found that the proportion of semi-natural habitats and field margins per farm to be 16 percent on organic farms, compared to 3.7 percent on conventional farms. It was also noted that in Germany organic farmers took specific measures to increase the habitat diversity (hedges, low-input orchards, ponds, corridors, habitat networks, wildlife refuges and devices) as well as to better connect farmed areas and the surrounding habitats.

A recent study, also in Switzerland, shows how important the combination and integration of semi-natural habitats with organic farms is for the conservation and enhancement of species diversity and abundance of beneficial arthropods. In this study organic agriculture had been compared with low-input integrated crop management within the framework of a paired farm survey. Also nearby semi-natural landscapes have been used to compare the effects of organic agriculture and low-input farming system on carabid beetle and spider fauna. Many endangered or rare species which are enhanced by semi-natural habitats and field margins were more abundant in organic arable fields than in the integrated managed fields. Also, agro-ecologically important carabid species and wolfspiders were found in higher numbers on the organic farms. This aspect indicates that the improvement of landscape infrastructure in combination with organic agriculture may be an important factor for the conservation and enhancement of species-rich communities on agricultural land. Furthermore, this approach has demonstrated a better natural pest control for a rape pest (the rape pollen beetle) in Germany and for the cherry oak aphid in Sweden.

Landscape Development and Organic Agriculture

Organic agriculture has the potential for a positive landscape development because it itself need a rich and diverse landscape infrastructure with semi-natural habitats. By studying the whole farm, or even the whole landscape within which the farm operates, researchers are trying to find ways to characterize the benefits of organic agriculture to the ecosystem. A research project funded by the European Union found that the diversity of landscape and farming systems was greater in organic farms, regarding land use types, crops, livestock, plantings (hedges, solitary shrubs, trees) and flora. In terms of landscape diversity, the organic types of agriculture have a good potential for positive contributions to a sustainable agrarian landscape.

Protected Areas

Protected areas, as defined by the World Commission on Protected Areas, are areas especially dedicated to the protection and maintenance of biological diversity. Protected areas are not always strict nature reserves. Instead, they can fulfil many functions alongside biodiversity conservation. Because of its biodiversity benefits, organic agriculture offers an important agricultural management option in several protected area categories.

In Italy for instance, the Italian Association for Organic Agriculture (AIAB) project on "Organic Agriculture and Agro-ecology in Regional Parks" has been working with the regional park authorities in the Emilia-Romagna region to promote organic agriculture in relation to the regional agri-environment programme. During the first two years of activity (1996-1997) there was an increase in the rate of adoption of the regional agri-environment programme, particularly of organic agriculture, by farmers in the park and the buffer zone. Between 1996 and 1997, 113 farms in the area applied for organic certification, compared with only 73 between 1994 and 1996.

There are further examples from Europe and the United States where organic agriculture is promoted in the biosphere reserves. The Estancia Itabo in Paraguay embodies a protected area of 5 000 hectares of high quality Interior Atlantic Forest. The Estancia is a good example of sustainable rainforest use in line with the conservation goals of the protected area by cultivating organic yerba mate and the heart of palm Euterpe edulis.

Buffer Zones

A secondary and closely connected link between ecosystem diversity, protected area management and organic agriculture is in the buffer zones (the region near the border of a protected area). Buffer zones are by their nature areas where land management aims to help maintain the integrity of the ecosystem of the core protected area. Where agriculture is the dominant land-use in buffer zones, the detrimental effects of farming systems can be reduced by conversion to organic systems.

The use of organic agriculture in protected buffer zones has been explored in the Meso-American Biological Corridor, a complex of protected areas and sustainable management stretching over seven countries. The initiative envisages a range of sustainable land uses within the buffer zones and linking areas, including certified forest management and organic agriculture.

In Peru, organic agriculture practices are promoted in the buffer zone of the Ampay Forest Sanctuary. Thanks to information measures, the awareness of the population about the biological importance of the sanctuary was increased, and as a result the area has converted into a high-priority conservation area for ecotourism, governmental and private investment in infrastructure, promotion and conservation of natural resources.

Where agriculture is the dominant land-use in buffer zones, the detrimental effects of farming systems can be reduced by conversion to organic systems.

Support for Organic Agriculture for its Biodiversity Benefits

There is increasing recognition among nature conservation organizations that many species interact with agricultural systems, even if their primary habitat is in natural areas. The management of these agricultural systems can, thus, dramatically affect overall levels of biodiversity, as well as the success of particular species. The German conservation organization BUND claims that traditional concepts of nature protection have failed because conservation goals have not been achieved, even in protected areas, due to agricultural intensification in the surrounding areas. Many conservation organizations are therefore calling for a general extensification of agricultural land use and promote organic agriculture for that reason.

In 1999, a joint workshop in Vignola, Italy, organized by The World Conservation Union (IUCN) and IFOAM together with the World Wide Fund for Nature (WWF), was held to exchange ideas and information on organic agriculture and biodiversity. A joint action plan for both the nature conservation and organic movements was drawn up in the so-called Vignola Declaration.

In both developed and developing countries, one of the driving forces behind the growth of organic agriculture is the nature conservation agenda. Though less significant than market prospects and development prospects, a recent literature survey highlights a number of examples where nature conservation organizations are working closely with local farmers who live in, or close to, areas of significant nature conservation interest.

In the last twenty years, governments also have been gradually moving towards policies aiming at encouraging the links between organic agriculture and biodiversity conservation.

In the European Union, environmental concerns have gained a more prominent place in agricultural policy. The most notable contribution of the Common Agricultural Policy of the European Union towards more environmentally sustainable systems has been the introduction of the agri-environment programmes in 1993, which now continue under Agenda 2000. The impetus for some of these initiatives comes from the Convention on Biological Diversity.

The implementation of agri-environmental measures over the whole of the European Union and also other European countries is the core of the European Community's environmental strategy. So far, organic agriculture has played a central role in most countries' national agri-environment policy. The main reason for this policy support has been the perceived positive environmental effects of organic agriculture. Also in other parts of the world, governments and development agencies are beginning to promote organic agriculture, especially in protected areas, because of its biodiversity benefits.

Organic agriculture brings several environmental benefits but the question arises on the threshold for converting land to organic management. One approach to determining this is to examine the benefits obtained from moving away from conventional agriculture. One estimate of the negative environmental externalities from conventional agriculture in the United States is derived from Pimentel et al. who estimated total indirect private and external (off-farm) costs of the use of synthetic pesticides at US $12.1 billion annually: this included the private input costs of US $4 billion and US $5 billion in environmental and health costs. Farmers had a US $12 billion incentive to use these pesticides. By comparison, the private benefits captured by the farmers were calculated at US $16 billion but in their decision they were ignoring the costs of US $8.1 billion imposed on the rest of society. These rather dated estimates indicate the nature and magnitude of the problem.

Redman estimated the costs of cleaning drinking water in the United Kingdom in excess of US $2 billion in initial investments plus US $240 million annually. These costs relate to pesticides and nitrate pollution. So far, organic agriculture has played a central role in most countries' national agri-environment policy.

Examples of environmental externalities should be used with caution because of the difficulties of assessing environmental damage and valuing the benefits of avoiding it. By their nature, there are no markets to value environmental externalities. One preventive approach to this problem is to set environmental standards and design measures that can best achieve these standards.

Another approach to estimating costs is to look at the costs of compliance to environmental regulations. A study comparing the impacts of environmental standards on agriculture in Australia, Canada, the European Union, New Zealand and the United States concludes that the compliance costs would normally be less than 3 or 4 percent of gross revenue, although intensive livestock industries face the greatest external costs.

Agricultural-related environmental problems relate chiefly to water quality and intensive livestock operations. To illustrate this point, research using the United Kingdom Government data suggests that the costs of meeting the Nitrates Directive in England and Wales reduced net farm income by 2.4 percent in the poultry sector, 11.3 percent in the pig industry, 4.5 for beef and 9.6 on dairy farms. Most of the cost relates to capital expenditure to adequately handle manure. Animal welfare regulations for poultry and pigs are estimated to reduce net farm income by 38 and 6 percent respectively. Crops do not face the same compliance costs. The implications of these estimates are that perhaps environmental concerns do not justify the abandonment of pesticides and synthetic fertilizers that a widespread shift to organic agriculture would entail. However, this approach ignores the environmental damage that occurs even though the regulations are being met.

There is some evidence that consumers who have recently entered the organic market are less motivated by ecological issues and are more interested by organic food as a safe product. What is interesting is that sales of organic products have increased markedly as a result of recent food scares, suggesting that new consumers are concerned about the product itself as well as, or perhaps rather than, the environment in which it is produced. However, the increases in sales have not related to the particular products (that is, beef) to which the scares applied. This distinction between food safety and environmental concerns has important implications for future production and policies.

CARBON FARMING

Carbon farming is a name for a variety of agricultural methods aimed at sequestering atmospheric carbon into the soil. Increasing the carbon content of soil can aid plant growth, increase soil organic matter (improving agricultural yield), improve soil water retention capacity, and reduce fertilizer use (and the accompanying emissions of greenhouse gas nitrous oxide (N_2O). As of 2016, variants of carbon farming affected hundreds of millions of hectares globally, of the nearly 5,000,000,000 hectares (1.2×10^{10} acres) of world farmland. Soils can contain up to five per cent carbon by weight, including decomposing plant and animal matter and biochar.

Potential sequestration alternatives to carbon farming include scrubbing the air with machines; fertilizing the oceans to prompt algal blooms that after death carry carbon to the sea bottom; storing the carbon dioxide emitted by electricity generation; and crushing and spreading types of rock such as basalt that absorb atmospheric carbon. Land management techniques that can be combined with farming include planting/restoring forests, burying biochar produced by anaerobically converted biomass and restoring wetlands. Coal beds are the remains of marshes and peatlands.

Soil Carbon

Traditionally, soil carbon was thought to accumulate when decaying organic matter was physically mixed with soil. More recently, the role of living plants has been emphasized. Small roots die and decay while the plant is alive, depositing carbon below the surface. Further, as plants grow, their roots inject some carbon into the soil, feeding mycorrhiza. An estimated 12,000 miles of their hyphae live in every square meter of healthy soil.

Techniques

At least thirty-two Natural Resource Conservation Service (NRCS) practices improve soil health and sequester carbon, along with important co-benefits: increased water retention, hydrological function, biodiversity and resilience. Approved practices may make farmers eligible for federal funds. Not all carbon farming techniques (e.g., composting) have been recommended. Carbon farming may consider related issues such as groundwater and surface water degradation.

Tilling

Carbon farming minimizes disruption to soils over the planting/growing/harvest cycle. Tillage is avoided using seed drills or similar techniques. Livestock can trample and eat the remains of a harvested field.

Livestock Grazing

Grazing livestock sequesters carbon when the animal eats the grass, causing its roots to release carbon into the soil. However, these animals also produce significant methane, potentially offsetting the benefits of sequestration. Livestock is regularly rotated through multiple paddocks (as often as daily) allowing the paddocks to rest/recover between grazing periods. This pattern produces stable grasslands with significant fodder. Annual grasses have shallower roots and die once they're grazed. Rotational grazing led to the replace of the annuals perennials with deeper roots, which can recover after limited grazing. By contrast, allowing animals to range over a large area for an extended period can destroy the grassland.

Silvopasture

Silvopasture involves grazing livestock under tree cover, with trees separated enough to allow adequate sunlight to reach the ground. For example, a farm in Mexico planted native trees on 22 hectares (54 acres) grazing cattle on the paddock. This evolved into a successful organic dairy. The operation became a subsistence farm, earning income from consulting/training others rather than from crop production.

Organic Mulch

Mulching covers the soil around plants with a mulch of wood chips or straw. Alternatively, crop residue can be left in place to enter the soil as it decomposes.

Compost

Compost sequesters carbon in a stable (not easily accessed) form. Carbon farmers dust it over the soil surface without tilling. A 2013, study found that a single compost application significantly and durably increased grassland carbon storage by 25-70%. The continuation sequestration likely came from increased water-holding and "fertilization" by compost decomposition. Both factors support increased productivity. Both tested sites showed large increases in grassland productivity: a forage increase of 78% in the drier valley site, while the wetter coastal site averaged an increase of 42%. CH_4 and N_2O and emissions did not increase significantly. Methane fluxes were negligible. Soil N_2O emissions from temperate grasslands amended with chemical fertilizers and manures were orders of magnitude higher. Another study found that grasslands treated with a single treatment 0.5" of commercial compost began absorbing carbon at an annual rate of nearly 1.5 tons/acre and continued to do so in subsequent years.

Cover Crops

With row crops such as corn and wheat, fast-growing ground cover is grown between the stalks (e.g., clover or vetch). They protect the soil from carbon loss through the winter and may be planted together with cash crops to compensate for carbon lost when those crops are harvested. Forage crops such as grasses, clovers and alfalfa develop extensive root systems that can become soil organic matter. Crops with poor root systems (corn, soybeans) do not increase organic matter in the soil.

Bamboo

Bamboo stores large amounts of carbon in the soil and in its superstructure.

Hybrids

Perennial crops offer the highest potential to sequester carbon, when grown in multilayered systems. These systems can be challenging and require farmers to cultivate potentially unfamiliar crops. One system uses perennial staple crops that grow on trees that are analogs to maize and beans, or vines, palms and herbaceous perennials.

Conventional Agriculture

Plowing splits soil aggregates and allows microorganisms to consume their organic compounds. The increased microbial activity releases nutrients, initially boosting yield. Thereafter the loss of structure reduces soil's ability to hold water and resist erosion, thereby reducing yield.

SUSTAINABLE GARDENING

Sustainable gardening includes the more specific sustainable landscapes, sustainable landscape design, sustainable landscaping, sustainable landscape architecture, resulting in sustainable sites. It comprises a disparate group of horticultural interests that can share the aims and objectives associated with the international post-1980s sustainable development and sustainability programs developed to address the fact that humans are now using natural biophysical resources faster than they can be replenished by nature.

Included within this compass are those home gardeners, and members of the landscape and nursery industries, and municipal authorities, that integrate environmental, social, and economic factors to create a more sustainable future. Organic gardening and the use of native plants are integral to sustainable gardening.

A water collector.

After the establishment of sustainable agriculture in the early 1980s it was some time before the emergence of Sustainable Horticulture (as sustainable production horticulture) at the International Society of Horticultural Science's First International Symposium on Sustainability in Horticulture held at the International Horticultural Congress in Toronto in 2002. This symposium produced "conclusions on Sustainability in Horticulture and a Declaration for the 21st Century". The principles and objectives outlined at this conference were discussed in more practical terms at the following conference at Seoul in 2006.

Many of the eco-friendly principles and ideas espoused by sustainable gardens, landscapes and sites perpetuate sustainable practices established as a reaction to resource-intensive industrial agriculture. These practices were established as movements for self-sufficiency and small-scale farming based on a holistic systems approach and ecological principles. Included here would be: biodynamic agriculture, no-till farming, agroecology, Fukuoka farming, forest gardening, organic gardening and others. On a larger scale there is the more recent "whole farm planning" which was established in 1995, and ecoagriculture established in 2000, and other variants of sustainable agricultural systems. Perhaps the most influential of these approaches is permaculture, established by Australians Bill Mollison and David Holmgren as both a design system and a loosely defined philosophy or lifestyle ethic. Permaculture shares many principles and practices of the above but not the broad philosophical base as indicated by the title of the 2002 publication Permaculture, principles and pathways beyond sustainability. The application of sustainability principles to the horticultural sphere has now become broadly accepted in commerce and academia.

Principles and Concepts

Managing global biophysical cycles and ecosystem services for the benefit of humans, other organisms and future generations has now become a global human responsibility. The method of applying sustainability to gardens, landscapes and sites is still under development and varies somewhat according to the context under consideration. However, there are a number of basic and common underlying biological and operational principles and practices in the sustainable sites literature.

Biological Principles

Sustainable management of man-made landscapes emulates the natural processes that sustain the biosphere and its ecosystems. First and foremost is the harnessing the energy of the Sun and the cycling of materials thereby minimising waste and energy use.

Running within, and dependent on, the natural economy there is the production and consumption of goods and services in the "human economy" which has now significantly altered, in a detrimental way, natural biogeochemical cycles (notable here are the water cycle), carbon cycle and nitrogen cycle so sustainable practices maximise support for ecosystem services.

Native Plants

The use of native plants in a garden or landscape can both preserve and protect natural ecosystems, and reduce the amount of care and energy required to maintain a healthy garden or landscape. Native plants are adapted to the local climate and geology, and often require less maintenance than exotic species. Native plants also support populations of native birds, insects, and other animals that they coevolved with, thus promoting a healthy community of organisms.

Plants in a garden or maintained landscape often form a source population from which plants can colonize new areas. Avoiding the use of invasive species helps to prevent such plants from establishing new populations. Similarly, the use of native species can provide a valuable source to help these plants colonise new areas.

Some non-native species can form an ecological trap in which native species are lured into an environment that appears attractive but is poorly suited to them.

However, in Britain research has revealed that for many invertebrates - the majority of wild animals in most gardens - it is not just native plants which can sustain them. The findings were published in popular form in Ken Thompson's book 'No Nettles Required: The truth about wildlife gardening'. He confirms the approach which Chris Baines had promoted in 'How to Make a Wildlife Garden'.

Operational Principles

Enhancement of ecosystem services is encouraged throughout the lifecycle of any site by providing clear design, construction, (operations), and management criteria. To be sustainable over the long term requires environmental, social and economic demands are integrated to provide intergenerational equity by providing regenerative sustainable systems. Operational guidelines will link to and supplement existing guidelines for the built environment (supplementing existing green building and landscape guidelines), the wider environment, and they will include metrics (benchmarks, audits, criteria, indexes etc.) that give some measure of sustainability (a rating system) by clarifying what is sustainable or not sustainable or, more likely, what is more or less sustainable.

Scale

Impacts of a site can be assessed and measured over any spatio-temporal scale or context.

Direct and Indirect Environmental Impact

Impacts of a site may be direct by having direct measurable impacts on biodiversity and ecology at the site itself or indirect when impacts occur away from the site.

Site Principles

The following are some site principles for sustainable gardening:

- Do no harm,
- Use the precautionary principle,
- Design with nature and culture,
- Use a decision-making hierarchy of preservation, conservation, and regeneration,

- Provide regenerative systems as intergenerational equity,
- Support a living process,
- Use a system thinking approach,
- Use a collaborative and ethical approach,
- Maintain integrity in leadership and research,
- Foster environmental stewardship.

Compost heap.

Measuring Site Sustainability

One major feature distinguishing the approach of sustainable gardens, landscapes and sites from other similar enterprises is the quantification of site sustainability by establishing performance benchmarks. Because sustainability is such a broad and inclusive concept the environmental impacts of sites can be categorised in numerous ways depending on the purpose for which the figures are required. The process can include minimising negative environmental impacts and maximising positive impacts. As currently applied the environment is usually given priority over social and economic factors which may be added in or regarded as an inevitable and integral part of the management process. A home gardener is likely to use simpler metrics than a professional landscaper or ecologist.

Three methodologies for measuring site sustainability include BREEAM developed by the BRE organisation in the UK, Leed, developed in America and the Oxford 360 degree sustainability Index used in Oxford Park and developed by the Oxford Sustainable Group in Scandinavia.

The Sustainable Sites Initiative is producing recommendations for the American Landscape Industry. The standards and guidelines finally adopted will lead to a uniform national standard, which does not currently exist. Sustainable Sites is currently in the pilot program stage, and will formally introduce its first rating system by 2013. The U.S. Green Building Council supports the project and plans to adopt the Sustainable Sites metrics into future versions of its Leadership in Energy and Environmental Design Green Building Rating System.

PERMACULTURE

Permaculture is a set of design principles centered on whole systems thinking, simulating, or directly utilizing the patterns and resilient features observed in natural ecosystems. It uses these principles in a growing number of fields from regenerative agriculture, rewilding, and community resilience.

The term permaculture was coined by David Holmgren, then a graduate student at the Tasmanian College of Advanced Education's Department of Environmental Design, and Bill Mollison, senior lecturer in Environmental Psychology at University of Tasmania, in 1978. It originally meant "permanent agriculture", but was expanded to stand also for "permanent culture", since social aspects were integral to a truly sustainable system as inspired by Masanobu Fukuoka's natural farming philosophy.

It has many branches including ecological design, ecological engineering, regenerative design, environmental design, and construction. Permaculture also includes integrated water resources management that develops sustainable architecture, and regenerative and self-maintained habitat and agricultural systems modelled from natural ecosystems.

Mollison has said: "Permaculture is a philosophy of working with, rather than against nature; of protracted and thoughtful observation rather than protracted and thoughtless labor; and of looking at plants and animals in all their functions, rather than treating any area as a single product system."

The 12 principles of permaculture most commonly referred to are first described by David Holmgren in his book Permaculture: Principles and Pathways Beyond Sustainability. They include: Observe and Interact, Catch and Store Energy, Obtain a Yield, Apply Self Regulation and Accept Feedback, Use and Value Renewable Resources and Services, Produce No Waste, Design From Patterns to Details, Integrate Rather Than Segregate, Use Small and Slow Solutions, Use and Value Diversity, Use Edges and Value the Marginal, and Creatively Use and Respond to Change.

Core Tenets and Principles of Design

The three core tenets of permaculture are:

- Care for the earth: Provision for all life systems to continue and multiply. This is the first principle, because without a healthy earth, humans cannot flourish.
- Care for the people: Provision for people to access those resources necessary for their existence.
- Fair share: By governing our own needs, we can set resources aside to further the above principles. This includes returning waste back into the system to recycle into usefulness. The third ethic is referred to as Fair Share, which reflects that each of us should take no more than what we need before we reinvest the surplus.

Permaculture design emphasizes patterns of landscape, function, and species assemblies. It determines where these elements should be placed so they can provide maximum benefit to the local environment. Permaculture maximizes useful connections between components and synergy of the final design. The focus of permaculture, therefore, is not on each separate element, but rather on the relationships created among elements by the way they are placed together; the whole becomes greater than the sum of its parts. Permaculture design therefore seeks to minimize waste, human labor, and energy input by building systems, and maximizes benefits between design elements to achieve a high level of synergy. Permaculture designs evolve over time by taking into account these relationships and elements and can evolve into extremely complex systems that produce a high density of food and materials with minimal input.

The design principles, which are the conceptual foundation of permaculture, were derived from the science of systems ecology and study of pre-industrial examples of sustainable land use. Permaculture draws from several disciplines including organic farming, agroforestry, integrated farming, sustainable development, and applied ecology. Permaculture has been applied most commonly to the design of housing and landscaping, integrating techniques such as agroforestry, natural building, and rainwater harvesting within the context of permaculture design principles and theory.

Twelve Design Principles

Twelve Permaculture design principles articulated by David Holmgren in his Permaculture: Principles and Pathways Beyond Sustainability:

- Observe and interact: By taking time to engage with nature we can design solutions that suit our particular situation.
- Catch and store energy: By developing systems that collect resources at peak abundance, we can use them in times of need.
- Obtain a yield: Ensure that you are getting truly useful rewards as part of the work that you are doing.
- Apply self-regulation and accept feedback: We need to discourage inappropriate activity to ensure that systems can continue to function well.
- Use and value renewable resources and services: Make the best use of nature's abundance to reduce our consumptive behavior and dependence on non-renewable resources.
- Produce no waste: By valuing and making use of all the resources that are available to us, nothing goes to waste.
- Design from patterns to details: By stepping back, we can observe patterns in nature and society. These can form the backbone of our designs, with the details filled in as we go.
- Integrate rather than segregate: By putting the right things in the right place, relationships develop between those things and they work together to support each other.
- Use small and slow solutions: Small and slow systems are easier to maintain than big ones, making better use of local resources and producing more sustainable outcomes.
- Use and value diversity: Diversity reduces vulnerability to a variety of threats and takes advantage of the unique nature of the environment in which it resides.
- Use edges and value the marginal: The interface between things is where the most

interesting events take place. These are often the most valuable, diverse and productive elements in the system.

- Creatively use and respond to change: We can have a positive impact on inevitable change by carefully observing, and then intervening at the right time.

Layers

Suburban permaculture garden with different layers of vegetation.

Layers are one of the tools used to design functional ecosystems that are both sustainable and of direct benefit to humans. A mature ecosystem has a huge number of relationships between its component parts: trees, understory, ground cover, soil, fungi, insects, and animals. Because plants grow to different heights, a diverse community of life is able to grow in a relatively small space, as the vegetation occupies different layers. There are generally seven recognized layers in a food forest, although some practitioners also include fungi as an eighth layer.

1. The canopy: The tallest trees in the system. Large trees dominate but typically do not saturate the area, i.e. there exist patches barren of trees.
2. Understory layer: Trees that flourish in the dappled light under the canopy.
3. Shrub layer: A diverse layer of woody perennials of limited height. Includes most berry bushes.
4. Herbaceous layer: Plants in this layer die back to the ground every winter (if winters are cold enough, that is). They do not produce woody stems as the Shrub layer does. Many culinary and medicinal herbs are in this layer. A large variety of beneficial plants fall into this layer. May be annuals, biennials or perennials.
5. Soil surface/Groundcover: There is some overlap with the Herbaceous layer and the Groundcover layer; however plants in this layer grow much closer to the ground, grow densely to fill bare patches of soil, and often can tolerate some foot traffic. Cover crops retain soil and lessen erosion, along with green manures that add nutrients and organic matter to the soil, especially nitrogen.

6. Rhizosphere: Root layers within the soil. The major components of this layer are the soil and the organisms that live within it such as plant roots and rhizomes (including root crops such as potatoes and other edible tubers), fungi, insects, nematodes, worms, etc.
7. Vertical layer: Climbers or vines, such as runner beans and lima beans (vine varieties).

Guilds

A guild is a group of species within which each provides a unique set of diverse functions that work in conjunction or harmony. There are many forms of guilds, including guilds of plants with similar functions that could interchange within an ecosystem, but the most common perception is that of a mutual support guild. Mutual support guilds are groups of animals, insects, plants, etc. that work well together. Plants may be grown for food production, draw nutrients from deep in the soil through tap roots, are nitrogen-fixing legumes, attract beneficial insects, and repel harmful insects. When grouped together in a mutually beneficial arrangement, these plants form a guild.

Edge Effect

The edge effect in ecology is the effect of the juxtaposition, or placing contrasting environments on an ecosystem. Permaculturists argue that where vastly differing systems meet, there is an intense area of productivity and useful connections. An example of this is the coast; where the land and the sea meet, there is a particularly rich area that meets a disproportionate percentage of human and animal needs. This idea is played out in permacultural designs by using spirals in herb gardens, or creating ponds that have wavy undulating shorelines rather than a simple circle or oval (thereby increasing the amount of edge for a given area).

Zones

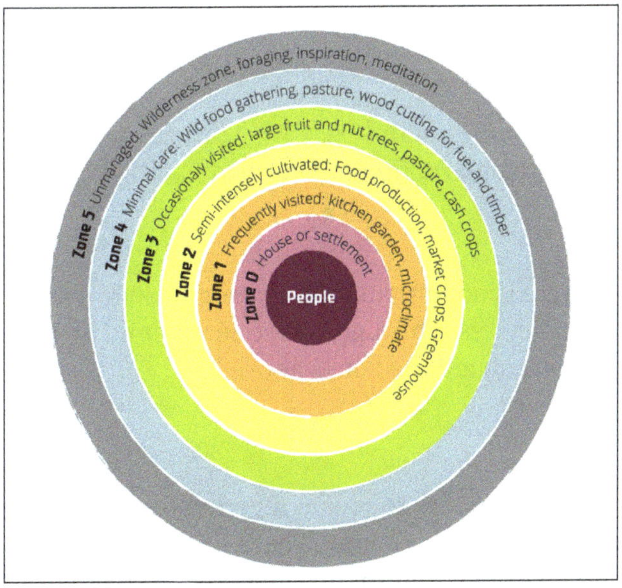

Permaculture zones 0–5.

Zones intelligently organize design elements in a human environment based on the frequency of human use and plant or animal needs. Frequently manipulated or harvested elements of the design are located close to the house in zones 1 and 2. Manipulated elements located further away are used less frequently. Zones are numbered from 0 to 5 based on positioning.

- Zone 0:

  The house, or home center. Here permaculture principles would be applied in terms of aiming to reduce energy and water needs, harnessing natural resources such as sunlight, and generally creating a harmonious, sustainable environment in which to live and work. Zone 0 is an informal designation, which is not specifically defined in Bill Mollison's book.

- Zone 1:

  The zone nearest to the house, the location for those elements in the system that require frequent attention, or that need to be visited often, such as salad crops, herb plants, soft fruit like strawberries or raspberries, greenhouse and cold frames, propagation area, worm compost bin for kitchen waste, etc. Raised beds are often used in zone 1 in urban areas.

- Zone 2:

  This area is used for siting perennial plants that require less frequent maintenance, such as occasional weed control or pruning, including currant bushes and orchards, pumpkins, sweet potato, etc. This would also be a good place for beehives, larger scale composting bins, etc.

- Zone 3:

  The area where main-crops are grown, both for domestic use and for trade purposes. After establishment, care and maintenance required are fairly minimal (provided mulches and similar things are used), such as watering or weed control maybe once a week.

- Zone 4:

  A semi-wild area. This zone is mainly used for forage and collecting wild food as well as production of timber for construction or firewood.

- Zone 5:

  A wilderness area. There is no human intervention in zone 5 apart from the observation of natural ecosystems and cycles. Through this zone we build up a natural reserve of bacteria, moulds and insects that can aid the zones above it.

People and Permaculture

Permaculture uses observation of nature to create regenerative systems, and the place where this has been most visible has been on the landscape. There has been a growing awareness though that firstly, there is the need to pay more attention to the peoplecare ethic, as it is often the dynamics of people that can interfere with projects, and secondly that the principles of permaculture can be used as effectively to create vibrant, healthy and productive people and communities as they have been in landscapes.

Domesticated Animals

Domesticated animals are often incorporated into site design, ensuring the efficiency and productivity of the system.Animals, domestic or wild are a critical component of any wild or designed sustainable ecosystem. Research indicates that without the animal's participation and contribution, ecological integrity is diminished or impossible. Some of the activities that contribute to the system include: foraging to cycle nutrients, clear fallen fruit, weed maintenance, spreading seeds, and pest maintenance. The nutrients are cycled by animals, transformed from their less digestible form (such as grass or twigs) into more nutrient-dense manure.

Several animals can be incorporated into a permaculture system, including cows, goats, chickens, geese, turkey, rabbits, and worms. A more specific explanation of how the animals can be used is seen in the chicken design. Chickens can be used to scratch over the soil, thus breaking down the top soil and using the fecal matter as manure creating a sustainable system. However, in the domestication of these animals, the complexity and elegance lie in an effectiveness and efficiency of the design, including factors like timing and habits to specific areas of a farm. For example, animals require daily attention in a way that is much more demanding than plants.

Vegan Permaculture

Vegan permaculture (also known as veganic permaculture, veganiculture, or vegaculture) avoids the use of domesticated animals. It is essentially the same as permaculture except for the addition of a fourth core value; "Animal Care." Zalan Glen, a raw vegan, proposes that *vegaculture* should emerge from permaculture in the same way veganism split from vegetarianism in the 1940s. Vegan permaculture recognizes the importance of free-living animals, not domesticated animals, to create a balanced ecosystem. Soil fertility is maintained by the use of green manures, cover crops, green wastes, composted vegetable matter in place of manure.

Common Practices

Agroforestry

Agroforestry with maize under trees.

Agroforestry is an integrated approach of permaculture, which uses the interactive benefits from combining trees and shrubs with crops or livestock. It combines agricultural and forestry technologies to create more diverse, productive, profitable, healthy and sustainable land-use systems. In agroforestry systems, trees or shrubs are intentionally used within agricultural systems, or non-timber forest products are cultured in forest settings.

Forest gardening is a term permaculturalists use to describe systems designed to mimic natural forests. Forest gardens, like other permaculture designs, incorporate processes and relationships that the designers understand to be valuable in natural ecosystems. The terms forest garden and food forest are used interchangeably in the permaculture literature. Numerous permaculturists are proponents of forest gardens, such as Graham Bell, Patrick Whitefield, Dave Jacke, Eric Toensmeier and Geoff Lawton.

Tree Gardens, such as Kandyan tree gardens, in South and Southeast Asia, are often hundreds of years old. It is not self-evident whether these tree gardens derived initially from experiences of cultivation and forestry, as is the case in agroforestry, or whether they derived from an understanding of forest ecosystems, as is the case for permaculture systems. Many studies of these systems, especially those that predate the term permaculture, consider these systems to be forms of agroforestry. Permaculturalists may obscure the distinction of permaculture and agroforestry when they include existing and ancient systems of polycropping as examples of food forests. Food forests and agroforestry are parallel approaches that sometimes lead to similar designs.

Hügelkultur

Hügelkultur is the practice of burying large volumes of wood to increase soil water retention. The porous structure of wood acts as a sponge when decomposing underground. During the rainy season, masses of buried wood can absorb enough water to sustain crops through the dry season. This technique has been used by permaculturalists Sepp Holzer, Toby Hemenway, Paul Wheaton, and Masanobu Fukuoka.

Natural Building

A natural building involves a range of building systems and materials that place major emphasis on sustainability. Ways of achieving sustainability through natural building focus on durability and the use of minimally processed, plentiful or renewable resources, as well as those that, while recycled or salvaged, produce healthy living environments and maintain indoor air quality.

The basis of natural building is the need to lessen the environmental impact of buildings and other supporting systems, without sacrificing comfort, health, or aesthetics. Natural building primarily utilizes abundantly available natural materials (e.g., clay, rock, sand, straw, wood, reeds), and draws heavily on traditional architectural strategies from various climates across the world. In addition to relying on natural building materials, the emphasis on the architectural design is heightened. The orientation of a building, the utilization of local climate and site conditions, the emphasis on natural ventilation through design, fundamentally lessen operational costs and positively impact the environment. Building compactly and minimizing the ecological footprint is common, as are on-site handling of energy acquisition, on-site water capture, alternate sewage treatment, and water reuse. Most materials are sourced regionally, locally, or even on-site. Straw bales, and various earthen masonry techniques such as adobe bricks, cob (or monolithic adobe), rammed earth and clay-straw infill are common choices for wall material. Roofing coverings often used include sod or "living" roofs, thatch, and wooden shakes or shingles. Rubble trench foundations are popular, as they do not require concrete; likewise, dry-stacked or lime mortared stem walls are common. Natural builders also regularly combine different wall systems in a single building, making best use of different materials' thermal or water resistant properties, for example, where they are most needed in the structure.

Rainwater Harvesting

Rainwater harvesting is the accumulating and storing of rainwater for reuse before it reaches the aquifer. It has been used to provide drinking water, water for livestock, water for irrigation, as well as other typical uses. Rainwater collected from the roofs of houses and local institutions can make an important contribution to the availability of drinking water. It can supplement the subsoil water level and increase urban greenery. Water collected from the ground, sometimes from areas which are especially prepared for this purpose, is called stormwater harvesting.

Greywater is wastewater generated from domestic activities such as laundry, dishwashing, and bathing, which can be recycled on-site for uses such as landscape irrigation and constructed wetlands. Greywater is largely sterile, but not potable (drinkable). Greywater differs from water from the toilets, which is designated sewage or blackwater to indicate it contains human waste. Blackwater is septic or otherwise toxic and cannot easily be reused. There are, however, continuing efforts to make use of blackwater or human waste. The most notable is for composting through a process known as humanure; a combination of the words human and manure. Additionally, the methane in humanure can be collected and used similar to natural gas as a fuel, such as for heating or cooking, and is commonly referred to as biogas. Biogas can be harvested from the human waste and the remainder still used as humanure. Some of the simplest forms of humanure use include a composting toilet or an outhouse or dry bog surrounded by trees that are heavy feeders which can be coppiced for wood fuel. This process eliminates the use of a standard toilet with plumbing.

Sheet Mulching

In agriculture and gardening, mulch is a protective cover placed over the soil. Any material or combination can be used as mulch, such as stones, leaves, cardboard, wood chips, gravel, etc., though in permaculture mulches of organic material are the most common because they perform more functions. These include absorbing rainfall, reducing evaporation, providing nutrients, increasing organic matter in the soil, feeding and creating habitat for soil organisms, suppressing weed growth and seed germination, moderating diurnal temperature swings, protecting against frost, and reducing erosion. Sheet mulching is an agricultural no-dig gardening technique that attempts to mimic natural processes occurring within forests. Sheet mulching mimics the leaf cover that is found on forest floors. When deployed properly and in combination with other Permacultural principles, it can generate healthy, productive and low maintenance ecosystems.

Sheet mulch serves as a "nutrient bank," storing the nutrients contained in organic matter and slowly making these nutrients available to plants as the organic matter slowly and naturally breaks down. It also improves the soil by attracting and feeding earthworms, slaters and many other soil micro-organisms, as well as adding humus. Earthworms "till" the soil, and their worm castings are among the best fertilizers and soil conditioners. Sheet mulching can be used to reduce or eliminate non-desired plants by starving them of light, and can be more advantageous than using herbicide or other methods of control.

Grazing

Grazing has long been blamed for much of the destruction we see in the environment. However, it has been shown that when grazing is modeled after nature, the opposite effect can be seen. Cell grazing is a system of grazing in which ruminant and non-ruminant herds or flocks are regularly and systematically moved to fresh pasture, range, or forest with the intent to maximize the quality and quantity of forage growth. Rotational grazing can be used with cattle, sheep, goats, pigs, chickens, rabbits, geese, turkeys, ducks, and other animals. Sepp Holzer and Joel Salatin have shown how the disturbance caused by the animals can be the spark needed to start ecological succession or prepare ground for planting. Allan Savory's holistic management technique has been likened to "a permaculture approach to rangeland management". One variation is conservation grazing, here the primary purpose of the animals is to benefit the environment and the animals can be, but are not necessarily, used for meat, milk or fiber. Sheep can replace lawn mowers. Goats and sheep can eat invasive plants.

Keyline Design

Keyline design is a technique for maximizing the beneficial use of water resources of a piece of land developed in Australia by farmer and engineer P. A. Yeomans. The *Keyline* refers to a specific topographic feature linked to water flow which is used in designing the drainage system of the site. The essential factor in this system, the Keyline, is a level or sloping line extending in both directions from a point or divides the two types of relationship, always in the same vertical interval, that a valley bears to its ridges.

Fruit Tree Management

Some proponents of permaculture advocate no, or limited, pruning. One advocate of this approach is Sepp Holzer who used the method in connection with Hügelkultur berms. He has successfully grown several varieties of fruiting trees at altitudes (approximately 9,000 feet (2,700 m)) far above their normal altitude, temperature, and snow load ranges. He notes that the Hügelkultur berms kept or generated enough heat to allow the roots to survive during alpine winter conditions. The point of having unpruned branches, he notes, was that the longer (more naturally formed) branches bend over under the snow load until they touched the ground, thus forming a natural arch against snow loads that would break a shorter, pruned, branch.

Masanobu Fukuoka, as part of early experiments on his family farm in Japan, experimented with no-pruning methods, noting that he ended up killing many fruit trees by simply letting them go, which made them become convoluted and tangled, and thus unhealthy. Then he realised this is the difference between natural-form fruit trees and the process of change of tree form that results from abandoning previously-pruned unnatural fruit trees. He concluded that the trees should be raised all their lives without pruning, so they form healthy and efficient branch patterns that follow their natural inclination. This is part of his implementation of the Tao-philosophy of Wú wéi translated in part as no-action (against nature), and he described it as no unnecessary pruning, nature farming or "do-nothing" farming, of fruit trees, distinct from non-intervention or literal no-pruning.

He ultimately achieved yields comparable to or exceeding standard/intensive practices of using pruning and chemical fertilisation.

Permaculture Action

Permaculture helps generate plans that are easy and cheap when it comes to production. Permaculture allows creativity and innovation in farming. The action of permaculture looks and all becoming consciously involved in the process of producing and ensuring abundant food nearby, problem on malnutrition caused by the world hunger problem will inevitably lessen. Permaculture principles in action are powerful forces to help right the environmental wrongs of the last two centuries.

POLYCULTURE

Polyculture providing useful within-field diversity: companion planting of carrots and onions. The onion smell puts off carrot root fly, while the smell of carrots puts off onion fly.

Polyculture is a form of agriculture in which more than one species is grown at the same time and place in imitation of the diversity of natural ecosystems. Polyculture is the opposite of monoculture, in which only members of one plant or animal species are cultivated together. Polyculture has traditionally been the most prevalent form of agriculture in most parts of the world and is growing in popularity today due to its environmental and health benefits. There are many types of polyculture including annual polycultures such as intercropping and cover cropping, permaculture, and integrated aquaculture. Polyculture is advantageous because of its ability to control pests, weeds, and disease without major chemical inputs. As such, polyculture is considered a sustainable form of agriculture. However, issues with crop yield and biological competition have caused many modern major industrial food producers to continue to rely on monoculture instead.

Common Practices

The kinds of plants that are grown, their spatial distribution, and the time that they spend growing together determines the specific type of polyculture that is implemented. There is no limit for the types of plants or animals that can be grown together to form a polyculture. The time overlaps between plants can be asymmetrical as well, with one plant depending on the other for longer than is reciprocated, often due to differences in life spans.

Annual Polycultures

Intercropping

When more than two crops are grown in complete spatial and temporal overlap with each other, it is referred to as intercropping. Intercropping is particularly useful in plots with limited land availability. Legumes are one of the most commonly intercropped crops, specifically legume-cereal mixtures. Legumes fix atmospheric nitrogen into the soil so that it is available for consumption by other plants in a process known as nitrogen fixation. The presence of legumes consequently eliminates the need for man-made nitrogen fertilizers in intercrops.

Cover Cropping

When a crop is grown alongside another plant that is not a crop, the combination is referred to as cover cropping. If the non-crop plant is a weed, the combination is called a weedy culture. Grasses and legumes are the most common cover crops. Cover crops are greatly beneficial as they can help prevent soil erosion, physically suppress weeds, improve surface water retention, and, in the case of legumes, provide nitrogen compounds as well.

Strip Cropping

Strip cropping is a form of polyculture that involves growing different plants in alternating rows. While strip cropping does not involve the complete intermixing of plant species, it still provides many of the same benefits such as preventing soil erosion and aiding with nutrient cycling.

Permaculture

Permacultures include polycultures of perennial plants such as legume-grass mixtures and wildflower mixtures, popular in Europe and more temperate climates. These increase soil fertility through nitrogen fixation, decrease soil erosion, regulate water consumption, and decrease the need for tillage thereby conserving soil nutrients. Permacultures require even less human intervention than other forms of polyculture because of lower harvest and tillage rates.

In many Latin American countries, agroforestry is a popular form of permaculture as well where trees and crops are grown together. Trees provide shade for the crops alongside organic matter and nutrients when they shed their leaves or fruits. The elaborate root systems of trees also help prevent soil erosion and increase the presence of microbes in the soil. In addition to benefiting crops, trees act as commodities themselves for use in paper, medicine, firewood, etc. Growing coffee plants alongside other tree species in Mexico is a common practice of agroforestry.

Coffee is a shade-loving crop, and is traditionally shade-grown. In India, it is often grown under a natural forest canopy, replacing the shrub layer. A different polyculture system is used for coffee in Mexico, where the Coffea bushes are grown under leguminous trees in the genus Inga.

A coffee farm where coffee plants are grown under several tree species in imitation of natural ecosystems. Trees provide valuable resources for the coffee plants such as shade, nutrients, and soil structure.

One approach to sustainability is to develop polyculture systems by breeding perennial crop varieties of traditional annual arable crops. Perennial crops require less tillage and often have longer roots than annual varieties, helping to reduce soil erosion and tolerate drought. Such varieties are being developed for rice, wheat, sorghum, pigeon pea, barley, and sunflowers. If these can be combined in polyculture with a leguminous cover crop such as alfalfa, fixation of nitrogen will be added to the system, reducing the need for fertilizer and pesticides.

Some traditional systems have combined polyculture with sustainability. In South-East Asia, rice-fish systems on rice paddies have raised freshwater fish as well as rice, producing a valuable additional crop and reducing eutrophication of neighbouring rivers. A variant in Indonesia combines rice, fish, ducks and water fern for a resilient and productive permaculture system; the ducks eat the weeds that would otherwise limit rice growth, saving labour and herbicides, while the duck and fish manure substitute for fertilizer.

Integrated Aquaculture

Integrated aquaculture is a form of aquaculture in which cultures of fish or shrimp are grown together with seaweed, shellfish, or micro-algae. Mono-species aquaculture, a form of aquaculture where only members of the species are grown together, poses several problems for farmers and the environment. The harvesting of seaweed crops in monocultures, for example, releases nitrates into the water and can lead to severe eutrophication as has occurred in the Venice Lagoon. In terms of seafood growth, the greatest problem in monocultures is the high cost of feed, which accounts for about half of production costs. However, more than half of seafood feed is shown to go to waste and can lead to further problems with excess nitrogen release and eutrophication or algal blooms of freshwater. Many technological approaches to lowering these harmful environmental effects such as bacterial bio-filters have proved to consume high levels of energy and be economically costly.

As such, many farmers have transitioned towards integrated aquaculture. In integrated aquaculture farms, plants serve a dual purpose, acting as food for the sea animals and as a water filtration device for the surrounding environment by absorbing nitrates and excess oxygen. Nutrients can be recycled between plants and animals, reducing the need for chemical nutrient supplements. Plants that are grown alongside seafood such as seaweed often hold significant commercial value by themselves, so incorporating them into already existing seafood monocultures increases economic value.

Functions

Pest Management

Pests are less predominant in polycultures than monocultures due to crop diversity. The lack of concentration of a single crop makes polycultures less appealing to pests who have a strong preference towards a specific crop. These specialized pests will often have more difficulty locating a favorable host plant inside of a polyculture than in a monoculture. If a pest has more generalized preferences, it will leave more quickly to other plants in the polyculture and as such have a lesser effect on any one plant. When pests are present in the nearby area, polycultures consequently experience lower yield loss than monocultures do in a theory known as the associational resistance hypothesis. Because polycultures mimic naturally diverse ecosystems, general pests are less likely to distinguish between polycultures and the surrounding environment as well. As such, pests travel more freely between the two environments, and have a relative smaller in presence in polycultures to begin with.

Because of the diversity of plants in a polyculture, natural enemies, or predators, of pests are often attracted to the polyculture alongside pests themselves as well. These natural enemies help suppress pest populations while doing no harm to the plants themselves.

Disease Control

Plant diseases are less predominant in polycultures than monocultures. The disease-diversity hypothesis states that a greater diversity of plants leads to a decreased severity of disease. Because different plants are susceptible to different diseases, if a disease negatively impacts one crop, it will not necessarily spread to another and so the overall impact is contained. However, the type of disease and the susceptibility of the specific plants inside the polyculture to a particular disease can vary greatly.

Weed Management

Both the density of crops and the diversity affect weed growth in polycultures. Having a greater density of plants reduces the available water, sunlight, and nutrient concentrations in the environment. Such a reduction is heightened with greater crop diversity as more potential resources are fully utilized. This level of competition makes polycultures particularly inhospitable to weeds.

When they do grow, weeds can help polycultures, assisting in pest management by attracting natural enemies of pests. They can also act as hosts to arthropods that are beneficial to other plants in the polyculture.

Advantages

Sustainability

Applying pesticides to crops in a monoculture: some of these pesticides will likely end up in water sources and the atmosphere where they can pose serious health and environmental harm. Polycultures attract fewer pests than monocultures and experience less yield loss even when pests are present, often eliminating the need for pesticides.

Because polycultures use methods of pest, disease, and weed control that do not rely on human intervention, they do not release pesticides into the environment. Fertilizer use is reduced as well, as diverse plants more fully share and use all available soil and atmospheric nutrients. As such, environmental impacts such as eutrophication of fresh water or the presence of excess atmospheric nitrogen are greatly reduced.

Other negative impacts of modern agriculture are similarly reduced. Excessive tillage occurs in most modern agricultural practices, but removes essential microbes and nutrients from the soil that are conserved in polycultures, especially permacultures. Because polyculture relies on natural systems of crop maintenance, farmers save money on machinery. By growing multiple plants or animals together in the same space, agricultural land, a critical resource worldwide, taking up 40% of the world's land area, is used more productively.

Polyculture increases local biodiversity. Increasing crop diversity can increase pollination in nearby environments, as diverse plants attract a broader array of pollinators. This is one example of reconciliation ecology, or accommodating biodiversity within human landscapes. This may also form part of a biological pest control program.

Human Health

The chemicals used in monoculture food production can be directly harmful to human health when released into the environment. Nitrogen is a chemical found in especially high concentrations in fertilizers. Nitrates from these fertilizers often become integrated in water sources due to agricultural runoff. The consumption of nitrates at high doses has been shown to lead to methemoglobinemia in infants.

Many of the crops consumed today are calorie-rich crops as well which can lead to illnesses such as obesity, hypertension, and type II diabetes. Because it encourages plant diversity, polyculture can help increase diet diversity by incorporating non-traditional foods into agriculture and people's diets.

Effectiveness

The effects of interspecific competition and intraspecific competition can cause great damage to plants in certain polycultures. In order for a polyculture to be effective, the diverse species that are a part of it must have distinct biological needs such as absorbing different nutrients or requiring different amounts of sunlight as stated by the competitive exclusion principle. Due to the large number of plant species that are cultivated by humans, finding and testing combinations of plants where interspecific and intraspecific competition do not significantly negatively affect the individual plants is extremely difficult. As such, for crops where historic polycultures do not exist, such a multiplicity makes the creation of new polycultures a significant issue.

Crop yield is also an issue in polycultures. While a polyculture produces more biomass overall than a monoculture, individual crops inside of the polyculture are not as prevalent. When there is a focal crop whose cultivation is especially important for a society a lower yield for a certain crop may pose food availability issues.

Similarly, while diseases and pests affect a polyculture less as a group, they do not necessarily have a decreased effect on a focal crop. If targeted by a specialized pest or disease, a focal crop in a polyculture will likely experience the same yield loss as its monoculture counterpart. Polyculture also often requires more labor.

SILVICULTURE

Silvics is the study of the life history and general characteristics of forest trees and crops, with particular reference to environmental factors, as the boisic for the practice of Silviculture.

Silviculture has even variously defined as:

- The art of producing and tending, a forest, the application of the knowledge of Silvics in the treatment of a forest, the theory and practice of controlling forest establishments, composition and growth.
- It is also defined as the art and science of cultivating forest crops or that branch of forestry which deals with the establishment, development care and regeneration of stands.

The subject of Silvicultural practice consists of the various treatments of forest stands that may be applied to maintain and enhance their productivity. The duties of the forester with respect to Silviculture are to analyze the natural and economic factors of significance on each stand under his care and to derive and apply treatment most appropriate to the objective of management. Silviculture occupies a position in forestry that is comparable or analogous to that of agronomy in agriculture. This is because it is concerned with the technical details of crop production. It is an applied science that rest ultimately upon this more fundamental natural and social sciences. In Silviculture, information of forest crops and technical procedures are developed for the scientific tending and reproducing of these crops. In a broad sense we thus have both the principles and practice of Silviculture.

The principle provides the scientific basis, while the practice is the application of the scientific basis.

The principles of Silviculture are concerned with the interpretation of forest vegetation as influenced by the environments which consist of actors of the habitat such as climate, soil and biotic factors. It views the forest as a complex structure with different biological units. It provides knowledge with the law governing production, on the capacity of forest stand and basis for little management around at forest development.

The practice of Silviculture deals with methods used for achieving the objectives. Thus, it can be called applied Silvics. It deals with the relationship between cultural method of cutting a stand and natural regeneration which is expected to follow the cutting.

- Also deals with the reduction and tending of stand e.g. thinning and pruning operations.
- Also concerned with methods of improving bole and methods of clearing and improving the quantity of the stand.
- Applicability and advantages of various methods.
- Comparison of various methods of collecting seed and other uses.

Place of Silviculture in Forest Resources

Forest resources are increasingly constituting a significant element in the national economies of many tropical countries. Unlike the past, when forests were taken for granted because they were found almost everywhere, the awareness has increased of the direct and for reaching influences of forest, as available forest area is continually diminished while the demand for forest goods and services soars.

The use of forest resources can be classified under:

- Major use e.g. wood as industrial raw materials.
- Intangible benefit such as recreation and watershed management.
- Minor produce e.g. food (fruits) useful plants tannins, gums etc.

The use of plant materials as raw materials now stands next in importance to their uses as food directly or indirectly. Wood is still the world's major structural material for building and other construction purposes. The Mahoganies of West African and the Meruntis of Malayan ran forest furnish the world with solid stocks and veneers for furniture. The mechanical/and chemical conversions of wood to pulp for the manufacture of paper and certain synthetic textiles consume enormous quantities of wood. Wood is aging treated to yield plastics, glues and special bonding agents. The distillation of wood yields a variety of valuable industrial chemicals such as methanol, acetic acids, turpentine, tanning materials and other products. In this age of energy shortages and oil politics the practice use and establishment of energy plantation or true plantation too harvest solar energy through bioconversion are being increasingly explored.

Besides supplying industrial raw materials, forest also have the added value of conservation for scenic purposes, stabilization of climate, maintenance of water supply and prevention of erosion. The dense, forest canopies form good protectors of soil against radiation and excessive diurnal temperature fluctuations. Vegetation generally serves as a good conserver of soil water by encouraging percolation and discouraging run-off. Its humus content also augments the water storage capacity of the soil for future use by booth plants and animals. Deforestation of upper parts of slopes often leads not only to increased evaporation of large quantities of water from the watershed but also alters the land use capabilities of the lower slopes trees also have an important function in breaking the forces of wind, thus preventing wind erosion. It has been shown by Haws that good wind-breaks induce relative calm up to twenty times the height of the trees on the leeward side and up to five times the height on the windward side such windbreaks help to increase the productivity of land by reducing the desiccating and abrasive of winds. Not only have these, forest provided food and habitat for wildlife. Wildlife management is almost impossible without some forms of forest management.

Modern forestry can then be defined as scientific management of forest resources for the continuous production of the important goods and services. However, unlike all other natural resources, forest resources are biologically renewable; they can grow and regrow after harvesting on the same site. Hence to work effectively, forest resources management must be biologically as well as economically sound.

Silviculture comes into forest resources management in the biological sector. Literally, it is understand to involve the formation of new or regeneration of old forests and caring for them until they are matured for the intended use. It is the science of how a forest crops can be produced naturally or artificially and cultured under the prevailing economic conditions to maturity for the projected use.

If the object of management is for cellulose production or as habitat or for recreational use, it is the duty of the Silviculturist to select those species and cultural techniques which will meet each objective. The Silviculturist is thus concerned not only with deciding what ought to be done but also the most economical way of achieving the various objectives. It must be noted that economic factors ultimately decide the silvicultural policy to be followed on any given area as the objective is usually to operate so as to maximize the value of benefits.

In recent times, the demand for land for other purposes and the general increase in cost of labour make intensive and efficient use of land available for forestry essential. It is apparent that only the most impressive evidence of prospective return may enable forestry areas of the tropics. The importance of sound training in silviculture to enable the forest establish and nurture his forest in the most profitable ways, cannot be over-emphasized.

The silculturist needs a clear understanding of the factors that govern the growth of forest trees, both as individual and as stands, the interactions between the plants and their environments and the likely consequences of change in parts of the forest ecosystem.

References

- Greene, Catherine. "USDA Economic Research Service - Organic Market Overview". Www.ers.usda.gov. Retrieved 19 April 2016
- Sustainable-agriculture-study-notes, agriculture: yourarticlelibrary.com, Retrieved 1 February, 2019
- Awada, L.; Lindwall, C.W.; Sonntag, B. (March 2014). "The development and adoption of conservation tillage systems on the Canadian Prairies". International Soil and Water Conservation Research. 2 (1): 47–65. Doi:10.1016/s2095-6339(15)30013-7. ISSN 2095-6339

- Agroforestry, science: britannica.com, Retrieved 2 March, 2019
- Holmgren, D. (2002). Permaculture, principles and pathways beyond sustainability. Holmgren Design Services, Hepburn, Australia. ISBN 0-646-41844-0
- Millennium Ecosystem Assessment (2005). Ecosystems and Human Well-being: Biodiversity Synthesis. Summary for Decision-makers. Pp.1-16. Washington, DC.: World Resources Institute. The full range of reports is available on the Millennium Ecosystem Assessment web site. [1]. Retrieved on: 2009-03-16

Permissions

All chapters in this book are published with permission under the Creative Commons Attribution Share Alike License or equivalent. Every chapter published in this book has been scrutinized by our experts. Their significance has been extensively debated. The topics covered herein carry significant information for a comprehensive understanding. They may even be implemented as practical applications or may be referred to as a beginning point for further studies.

We would like to thank the editorial team for lending their expertise to make the book truly unique. They have played a crucial role in the development of this book. Without their invaluable contributions this book wouldn't have been possible. They have made vital efforts to compile up to date information on the varied aspects of this subject to make this book a valuable addition to the collection of many professionals and students.

This book was conceptualized with the vision of imparting up-to-date and integrated information in this field. To ensure the same, a matchless editorial board was set up. Every individual on the board went through rigorous rounds of assessment to prove their worth. After which they invested a large part of their time researching and compiling the most relevant data for our readers.

The editorial board has been involved in producing this book since its inception. They have spent rigorous hours researching and exploring the diverse topics which have resulted in the successful publishing of this book. They have passed on their knowledge of decades through this book. To expedite this challenging task, the publisher supported the team at every step. A small team of assistant editors was also appointed to further simplify the editing procedure and attain best results for the readers.

Apart from the editorial board, the designing team has also invested a significant amount of their time in understanding the subject and creating the most relevant covers. They scrutinized every image to scout for the most suitable representation of the subject and create an appropriate cover for the book.

The publishing team has been an ardent support to the editorial, designing and production team. Their endless efforts to recruit the best for this project, has resulted in the accomplishment of this book. They are a veteran in the field of academics and their pool of knowledge is as vast as their experience in printing. Their expertise and guidance has proved useful at every step. Their uncompromising quality standards have made this book an exceptional effort. Their encouragement from time to time has been an inspiration for everyone.

The publisher and the editorial board hope that this book will prove to be a valuable piece of knowledge for students, practitioners and scholars across the globe.

Index

A
Abiotic Factors, 18, 33, 65, 131
Abiotic Stress, 65-66, 68-73, 77, 84, 88, 90-106, 108
Agri-environmental Schemes, 27, 33-38
Agricultural Management, 51, 56, 69, 183
Agroecosystem Analysis, 38-39
Alauda Arvensis, 24, 27
Algal Blooms, 24, 160, 186, 203
Amino Acid, 80, 82-83, 136
Ammonia, 44-46, 160
Atmospheric Nitrogen, 19, 44-46, 148, 202, 205
Azospirillum Brasilens, 77
Azotobacter, 45, 82, 109

B
Bacillus Amyloliquefaciens, 83, 87
Biological Nitrogen Fixation, 45
Biomass, 3, 23, 34, 42, 49, 56-57, 62, 76-80, 82, 84-85, 102, 108, 133, 152, 166, 175, 177, 180, 186, 206
Biotic Factors, 33, 65, 207
Biotic Stress, 66-67, 69, 91, 93-94, 96, 98-99, 102, 106, 108

C
Caffeic Acid, 73, 85
Carbon Cycle, 40, 48, 51, 189
Carbon Sequestration, 18-19, 26, 48, 52-53, 152, 155, 160
Cash Crops, 66, 141-142, 156, 187
Coleoptera, 28, 30, 130
Common Carp, 27-28
Crop Failure, 1, 142
Cupressus Macrocarpa, 25
Cyanobacteria, 46, 48

D
Decomposition, 19, 22, 41, 44, 48-50, 56, 165-166, 187
Deforestation, 52, 75, 208
Diptera, 28, 130
Drought Stress, 66, 76, 78-83, 87, 90, 94-96, 98, 108
Dry Matter, 49, 79-80

E
Ecological Recycling, 40, 42-43
Ecological Restoration, 23-24, 26-34
Eutrophication, 23, 52, 62, 160, 168, 203, 205
Exopolysaccharide, 77-78, 84

F
Fabaceae, 46, 155
Focal-species, 26-27
Fusarium Oxysporum, 80, 92, 134

G
Genetically Modified Organisms, 7, 73, 153, 159
Gibberellic Acid, 68, 73, 84
Glycine Max, 79, 106
Grassland, 19, 28-29, 31-32, 172, 174-175, 187
Green Manure, 46, 110, 153, 155-156, 165-166
Greenhouse Gas, 51-52, 75, 160, 186

H
Habitat Restoration, 26, 32, 34
Hydrogen Cyanide, 77-78

J
Jasmonic Acid, 67-68, 73, 86, 91, 93

L
Land Cover, 13-14, 23, 55
Land Degradation, 2, 12, 14, 51, 53, 62-63
Lantana Camara, 27-28

M
Manduca Sexta, 68, 91, 115
Medicago Truncatula, 81, 99
Mineral Nutrients, 40-42, 44
Mineralization, 48-50, 52, 165-166
Mulching, 2, 156, 166, 175, 187, 199

N
Native Vegetation, 24, 31
Nitrogen Fixation, 16, 21, 44-46, 76-77, 82, 84, 87, 102, 202
Nitrogenases, 45-46
Nucleic Acids, 45, 68
Nutrient Cycling, 16, 18-20, 27, 34-35, 40-41, 43-44, 52, 56, 62, 66, 177, 202

P
Phytohormone, 67, 84
Phytophthora Cinnamomi, 24
Pinus Radiata, 25
Pipistrellus Pipistrellus, 27-28
Plant Growth-promoting Rhizobacteria, 73, 75-76, 85
Plant Stress, 65-66, 88, 91, 97-98, 100-101, 105-106
Precision Farming, 2
Protozoa, 23, 42
Pseudomonas Fluorescens, 81, 83

R
Reactive Oxygen Species, 69-70, 93, 95, 97
Riparian Vegetation, 25-26
Root Nodule, 46-47
Rural Dieback, 24, 28

S
Salicylic Acid, 67-68, 91, 93, 109
Sodium Polyacrylate, 73, 82
Soil Carbon, 48, 51, 186
Soil Erosion, 13, 23, 33, 57, 63, 142, 149, 156, 162, 165-167, 174, 202-203
Soil Organic Matter, 6, 14, 48, 141, 162, 165-166, 186-187
Stomatal Closure, 66, 93, 97
Stress-tolerant Crop, 69, 88
Striga Hermonthica, 69, 89
Sustainable Agricultural, 9, 20, 24, 148, 189

V
Vegetal Detritus, 49-50

W
Water Cycle, 40, 189
Wind Erosion, 63, 70-71, 208

www.ingramcontent.com/pod-product-compliance
Lightning Source LLC
LaVergne TN
LVHW080652210926
848485LV00003B/366